S0-EKZ-908

First American edition published in 1993 by

Peter Bedrick Books
2112 Broadway
New York, NY 10023

Published by agreement with Oxford University Press, England

Library of Congress Cataloging-in-Publication Data
Goodman, Susan
Amazing biofacts/Susan Goodman: — 1st American ed.
Included index.
ISBN 0–87226–364–9 (hard)
ISBN 0–87226–256–1 (pbk)
1. Biology–Popular works. 2. Human biology–Popular works.
I. Title.
QH309.G68 1993
574–dc20 92–36376

Designed and typeset by Threefold Design
Printed in Hong Kong
10 9 8 7 6 5 4 3 2 1

CONTENTS

How manifold are Thy works, O Lord!
In wisdom hast Thou made them all;
The Earth is full of Thy creatures.

PSALM 104 VERSE 24

מָה־רַבּוּ מַעֲשֶׂיךָ ה׳
כֻּלָּם בְּחָכְמָה עָשִׂיתָ
מָלְאָה הָאָרֶץ קִנְיָנֶךָ

THE HUMAN BODY

CELLS

● In an adult human's body there are billions of cells. They make up the brain, heart, blood, skin and every other part of your body.

● The egg cell, which is the beginning of a baby, is one of the largest cells in the body and measures 0·2 mm (0·008 in) across, which is about the size of the dot on this 'i'. Brain cells which are the smallest cells in the body are only 0·005 mm (0·0002 in) across: about 40 of them will fit across the full-stop at the end of this sentence.

● Although there are many different kinds of cells they each have the same basic structure. A cell is surrounded by a cell wall which allows substances to move in and out of the cell. The cell is filled with a jelly-like substance called cytoplasm which contains everything the cell needs to stay alive. At the center of the cell is the nucleus, the control centre which provides all the instructions the cell needs to do its job properly. The detailed instructions are contained in complicated strings of chemicals called chromosomes.

● All the cells in our bodies, except egg cells and sperm cells, contain 46 chromosomes.

The 46 chromosomes from the body cell of a male. We know it is a male because of the X and Y chromosomes. Females have two X chromosomes and no Y chromosomes. The chromosomes are in pairs, so the genes carried by the chromosomes are in pairs too. Half of the genes are from the mother and half are from the father.

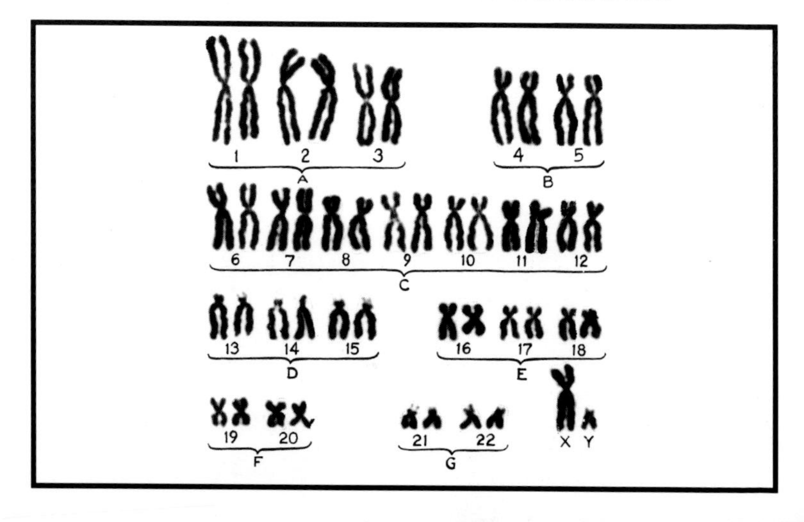

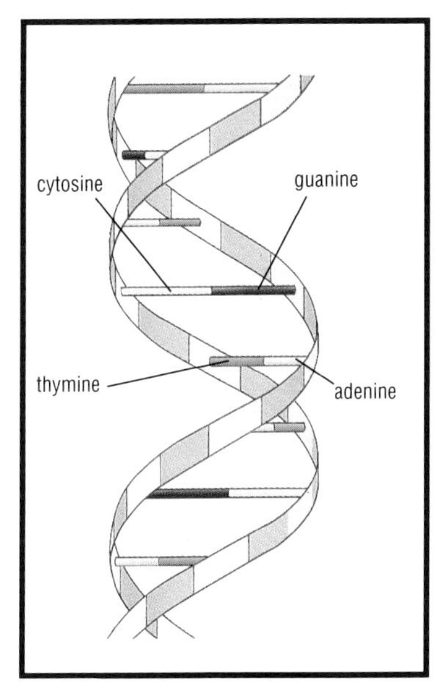

DNA is found in every cell in your body. It looks like a twisted ladder with each rung made up of one or two different chemical pairs. This long ladder consists of thousands of shorter ladders joined together. These are your genes.

Egg cells and sperm cells have only 23 chromosomes. When the egg is fertilized by the sperm these form 23 pairs; a total of 46 chromosomes.

● Chromosomes are made of a complicated chemical called deoxyribonucleic acid, or DNA for short. DNA is a very long chemical with a structure like a spiral ladder, made of small units called genes. It is the genes which carry the genetic information that determines exactly what we look like. The genes also control what job a particular cell does, ensuring that brain cells do the work of the brain rather than making blood like liver cells, or growing hairs like skin cells. But remember that all these cells contain exactly the same DNA inside their nucleus; it just depends which genes of the DNA are switched on. Nobody yet knows exactly what switches a particular gene into action.

● A DNA molecule in the nucleus of a cell is made up of between 100,000 and 10 million atoms and weighs 6 millionths of a millionth of a gram (0·000 000 000 006 g). It is this minute speck which contains all the genetic information we inherit from our parents, such as our sex, hair color, length of nose, and the likelihood of us getting certain diseases.

DID YOU KNOW?

If one cell was taken from everyone living on Earth and all the DNA was removed and weighed, it would weigh only 0.024 g (less than a thousandth of an ounce) in total.

IN THE BEGINNING

● The life of a baby begins inside the mother when a sperm from the father unites with one of her eggs to form a new cell. In about 9 months this cell will have divided many times and grown into a baby.

● Sperm look a little like tiny tadpoles. Their tails are about 0·3 mm (0·0012 in) long and move with a whip like motion so that the sperm swim towards the egg. About 200 million sperm enter the woman's vagina but only a few hundred arrive in the Fallopian tubes where the eggs are released. They have taken about an hour to swim there. Here, if they meet an egg just one sperm will enter and unite with it. This is called fertilization. The new cell moves down the Fallopian tube to the uterus where it becomes embedded, and usually remains until it is born a fully formed baby.

● Thirty hours after the egg is fertilized by a sperm it will divide into two cells. Twenty hours later these two cells divide producing a total of four cells. The cells carry on dividing: some will form the skin, others the brain, the heart and all the other different parts of our bodies.

IN THE WOMB

When a baby girl is born she has about 300,000 eggs already in her ovaries. About 400 of these will mature during her life and some of these may be fertilized and develop into her own babies.

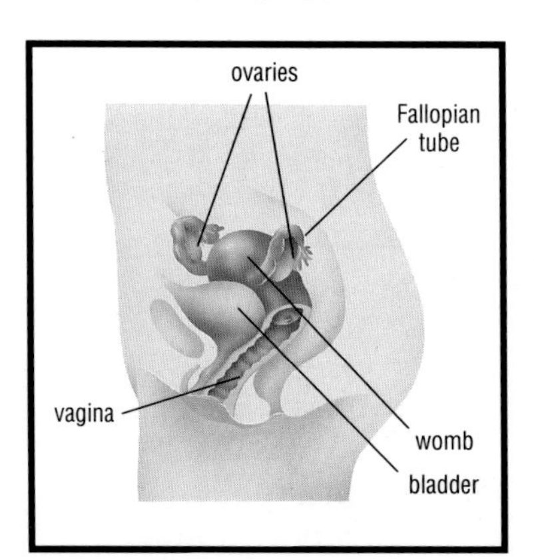

Female sex organs

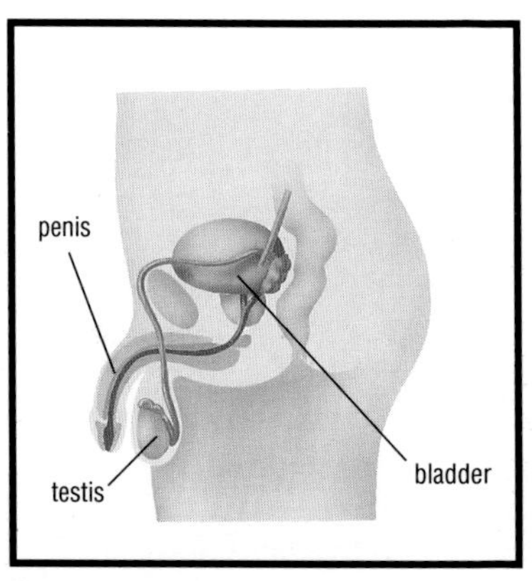

Male sex organs

THE DEVELOPING BABY

Week 1	Day 0-7	The fertilized egg is 0·2 mm (0·008 in) across.
Week 3	Day 15-21	Eyes and ears begin to form. The embryo is 2·5mm (0·1 in) long.
Week 4	Day 22-28	Heart starts beating. Tiny bumps appear where arms and legs will grow. There are about 10,000 cells. The embryo is 6·3 mm (¼ in) long.
Week 5	Day 29-35	Nose and stomach begin to form. The embryo is 13 mm (½ in) long.
	Day 31	Arms, hands and shoulders begin to take shape.
	Day 33	Outline of fingers appear.
Week 6	Stomach, intestines, liver, lungs and brain all developing. The embryo is 19 mm (¾ in) long. Tip of nose appears. Eyelids begin to form. Outline of toes can be seen.	
Week 7	Kidneys and liver start functioning. Ears almost complete.	
Week 9	There are definite palmprints and footprints. Nails begin to grow. The fetus is 4 cm (1½ in) long.	
Week 11-14	All the main organs are now developed, and the baby can frown, swallow and urinate. He or she now weighs 21 g (¾ oz) and is 7 cm (2 ¾ in) long.	
Week 15-18	Eyelashes and eyebrows begin to grow. The mother can probably feel her baby kicking for the first time.	
Week 19-22	If a baby is born now it is possible, with modern hospital facilities, that he or she might survive at 30 cm (12 in) long and weighing about 0·62 kg (1lb 6 oz).	
Week 23-26	Pictures of babies in the womb show them sucking their thumbs. Many babies born now can survive.	
Week 38	The baby continues to grow and put on weight. On average babies are born 266 days after the egg is fertilized. They weigh about 3·3 kg (7 lb 4 oz) and are 51 cm (20 in) long. They have just ½ a pint of blood but the heart pumps 340 litres (600 pints) through it a day.	

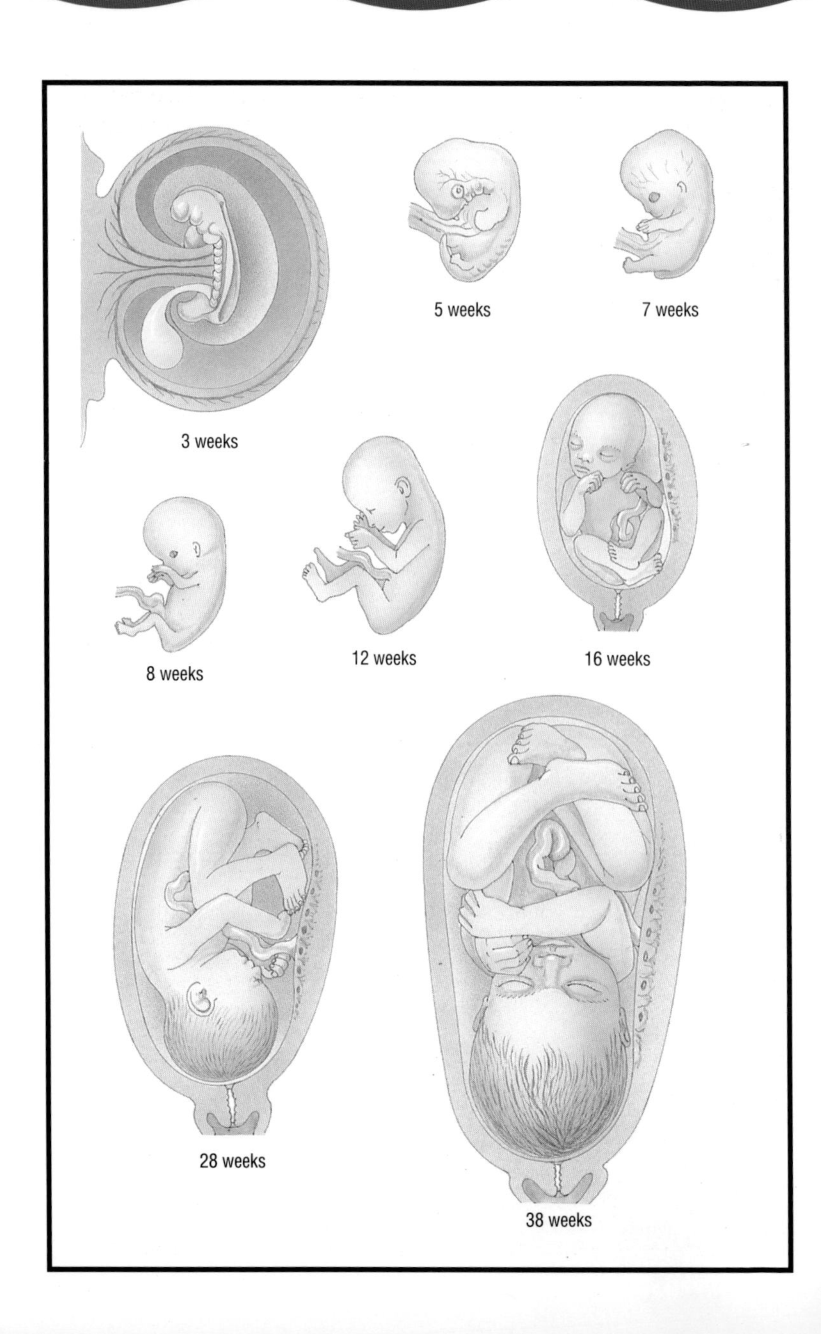
5 weeks
7 weeks
3 weeks
8 weeks
12 weeks
16 weeks
28 weeks
38 weeks

GROWTH AND DEVELOPMENT

● Humans are the only animal to give birth to a baby that remains so helpless for so long. It takes about eight months before babies can sit up without being supported, and to even begin trying to feed themselves. At a year old they take their first tottering steps.

● Our bodies continue developing and growing until we are about 20. Our mental abilities are usually thought to be at their best between the ages of 20 and 30 but most of us will continue to be mentally active until very old.

● There is no other animal like us. We may be slow beginners but what we can achieve in a lifetime is amazing.

● At birth babies have very little control over their muscles. But they do have some interesting reflex actions. These muscle movements happen automatically and most have disappeared by the time babies are a few weeks old.

Stepping reflex: Place babies with their feet on the ground and they will try and take steps.

Moro reflex: Move babies suddenly and they will throw their arms outwards.

Grasp reflex: Babies will close their fists firmly round anything placed in them. The grip is so tight that it can support the whole weight of the baby.

Rooting reflex: Stroke a baby's cheek and he or she will turn their head in that direction, pursing the lips ready to suck.

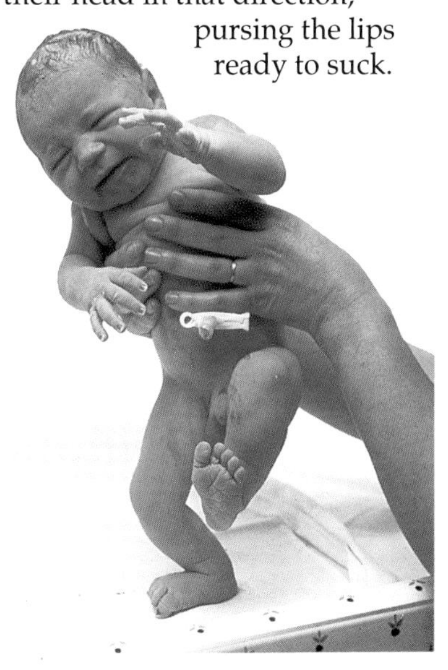

The development of the embryo and fetus in the womb. For the first eight weeks of life the term embryo is used, after that fetus is used until birth.

Stepping reflex in a newborn baby. This baby is only minutes old and the remainder of his umbilical cord has been clamped, where it has been cut, by a white clip.

TEETH

Our first teeth begin to arrive when we are about 7 months old. There are 20 teeth in this first set and they are called milk teeth.

Adult teeth

At about 6 years old adult teeth begin to push their way into the mouth and replace the milk teeth. It takes many years for the complete set of 32 adult teeth to come through. The last 4, the wisdom teeth, often do not appear until about age 25, and sometimes never emerge.

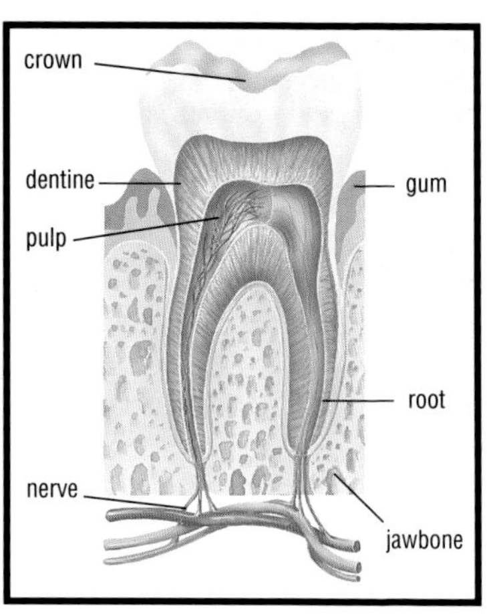

The structure of a molar. The crown is protected by a layer of enamel, the hardest substance in the human body.

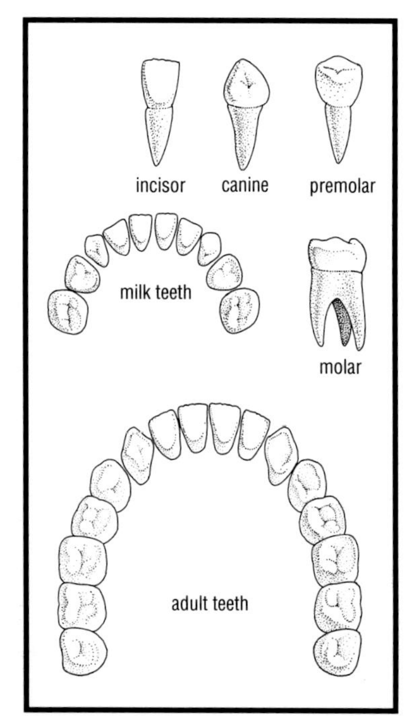

Tooth	average age at arrival (in months)
lower / upper } left incisors	7 ½
upper / lower } right incisors	9 ½
upper / lower } left premolars	15 ½
upper / lower } canines	19
lower / upper } right premolars	26 / 27

GROWTH

Chemicals which control how much we grow are produced in the pituitary gland in the brain. These growth hormones are

released into the blood stream during deep sleep.
The pituitary gland also controls the release of sex hormones. At the beginning of puberty these hormones produce development in a girl's ovaries and a boy's testes. They produce all the changes that make our bodies able to have babies.
Puberty usually follows a sudden increase in growth. It comes earlier for girls than boys, so often 12–14 year old girls are taller and heavier than boys of the same age. The first signs of puberty for girls occurs at between 10 and 14 years old and for boys between 12 and 15 years. It takes several years for the sex organs to reach maturity.

- Some girls continue growing until about 20 and some boys are not fully grown until they are 25 years old.

DID YOU KNOW ?

The tallest person in the world measured 2·7 m (8 ft 10 in) tall. The shortest person was 70 cm (27 ½ in) tall.

HOW TALL WILL YOU BE?

At nine years old an average, healthy boy is three-quarters of his adult height. A girl reaches three-quarters of her adult height when she is seven and a half years old.
A boy's height at two years old and a girl's at 18 months are about half what they will be at 18 years.

Age years	Heart rate beats per min	Heart weight gm (oz)	Brain weight kg (lbs)	breaths per min	Breathing volume or air (cc)
newborn	140	24 (0·85)	0·35 (0·75)	30-80	20
1	115	45 (1·6)	0·91 (2)	20-40	48
2	110				90
7	95	99 (3 ½)	1·2 (2·7)		
20	82	300 (10 ½)	1·4 (3)	15-20	500

Table of some of the changes as we grow up.

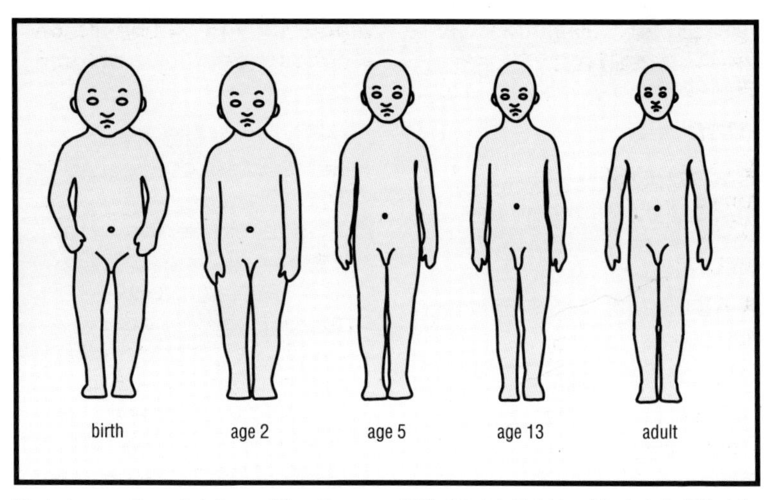

The body proportions of a baby are different to those of an adult. In a baby the head makes up 25% of the total height and the brain is 25% of its adult size.

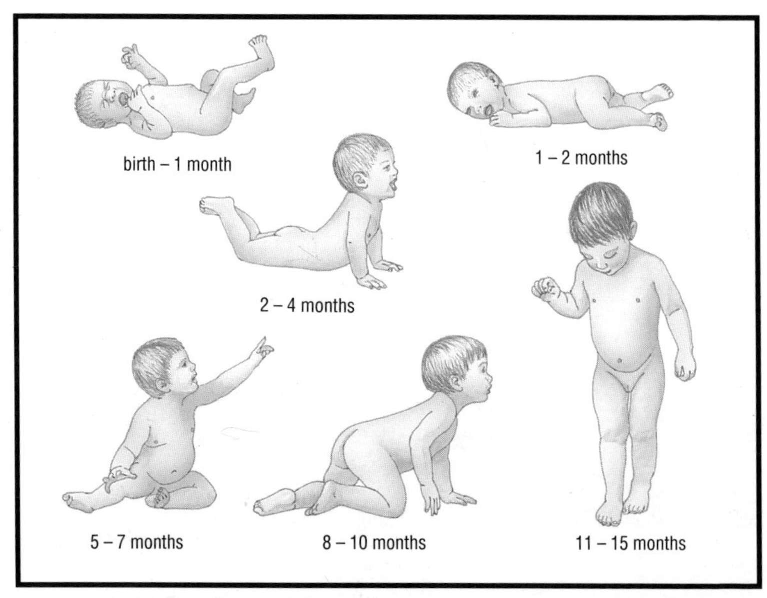

A baby's development during the first year. At birth a baby cannot even support her own head. She cannot sit unaided till about 6 months. Crawling begins at around the eighth month and walking at about a year.

FOOD AND DIGESTION

FOOD

- On average a man eats 50 tons of food and 50,000 litres (13,000 gal) of liquid during his lifetime.

- All the food we eat falls into five basic categories:

Carbohydrates such as sugars and starches give us energy. They are found in potatoes, bread and jam, among other things. The amount of energy produced by the food is measured in kilojoules (kJ) or kilocalories (calories). When resting you use about 400 kJ. Here is the extra energy you need if you do any of the following activities for an hour:

Activity	Calories/kcal	Kilojoules
swimming	600	2500
walking fast	240	1000
cycling	175	730
walking slowly	170	700
standing	40	170
writing	30	125

If you eat more carbohydrates than you need they are stored as fat.

Proteins are found in eggs, meat, fish, cereals and beans. During digestion the proteins are broken down into small units called amino acids. The body then rearranges these amino acids to form the particular proteins we need. There are only 20 essential amino acids but these can be built into millions of different proteins. They are part of every cell in your body.

Fats are stored as a layer of insulation under the skin. Fat produces twice as much energy as carbohydrate. About a quarter of the energy we use each day comes from fat. Sources of fat include butter, nuts, fatty meat and vegetable oil.

Minerals describes a whole collection of elements including: calcium and phosphorus, needed to build bones and teeth; sodium and potassium to keep nerves functioning properly; and tiny quantities of copper, cobalt, manganese and molybdenum. In an adult these last four elements weigh less than a gram, but they are all absolutely essential for health.

Vitamins are essential to your health but are only needed in tiny quantities. For example you need about 0·001 g (0·0003 oz) of vitamin B1, thiamine, each day. It only amounts to 28 g (1 oz) in your whole lifetime but without it your heart and nervous system will not function properly. About 12 vitamins are essential to health.

● A lifetime's supply of all the vitamins you need only weighs about 250 g (8 oz).

Vitamin	Food source	If lacking in diet
A	carrots, liver green vegetables, milk	growth slows down; more likely to get ill; unable to see well at night.
B group	yeast, potatoes, bread, meat, nuts, eggs	growth and development slow down; general health poor.
C	citrus fruit, potatoes and green vegetables	teeth and gums become unhealthy; skin sores develop.
D	margarine, milk, eggs, fish oils, sunshine on skin	weak bones and teeth
E	wholemeal bread, brown rice, butter, and green vegetables	cell growth and wound healing are believed to be affected.
K	green vegetables, liver, tomatoes	blood does not clot

FOOD TABLE

Food	Amount	Protein gm	Carbohydrate gm	Fat gm	Calories kcal	Energy kJ
Bread	3½ oz	7·3	40	2·4	200	833
Breakfast cereal	3½ oz	6·7	83	1·7	333	1400
Hamburger	3½ oz	16	2	22	255	1070
Fish sticks (x4)	3½ oz	16	20	15	240	1000
Fries	3½ oz	2·6	26	5	150	630
Baked beans	3½ oz	5	11	0·4	65	270
Spaghetti	3½ oz	2	14	0·4	60	250
Pizza	3½ oz	15	37	10	280	1200
Cheddar cheese	3½ oz	25	-	34	400	1680
Rice	3½ oz	6	52	0·4	225	940
Hazelnuts	3½ oz	16	4	65	660	2760
Ice cream	3½ oz	4	21	10	190	800
Chocolate	3½ oz	9	60	25	500	2090
Chips	3½ oz	7	43	36	500	2090
Chocolate cookie (x 5)	3½ oz	3	45	15	300	1250

THE ENERGY AND PROTEIN WE NEED:

Age years	Body weight lbs	Calories kcal	Energy kJ	Protein gm
Males				
9-12	70 lbs.	2500	10,500	63
12-15	100 lbs.	2800	11,700	70
15-18	135 lbs.	3000	12,600	75
18-35	144 lbs.	3000	12,600	68
Females				
9-12	73 lbs.	2300	9,600	58
12-15	108 lbs.	2300	9,600	58
15-18	124 lbs.	2300	9,600	58
18-35	121 lbs.	2200	9,200	55

DIGESTION

Your digestive system is basically a tube 10 m (33 ft) long which begins in the mouth and ends at the anus. This tube squashes and mixes food with acids and chemicals until the big bite in your mouth is reduced into small molecules which can pass through the wall of the gut and into your blood stream. Waste material passes out at the end of the intestines. It takes about 24 hours for food to complete this journey.

Mouth: Teeth pulp the food and the tongue mixes it with saliva. The chemicals in saliva are the first stage in breaking food down into small molecules. An adult produces over 1 ½ litres (2 ½ pints) of saliva every 24 hours.

Esophagus: This tube connects the mouth to the stomach. Food will pass down this tube

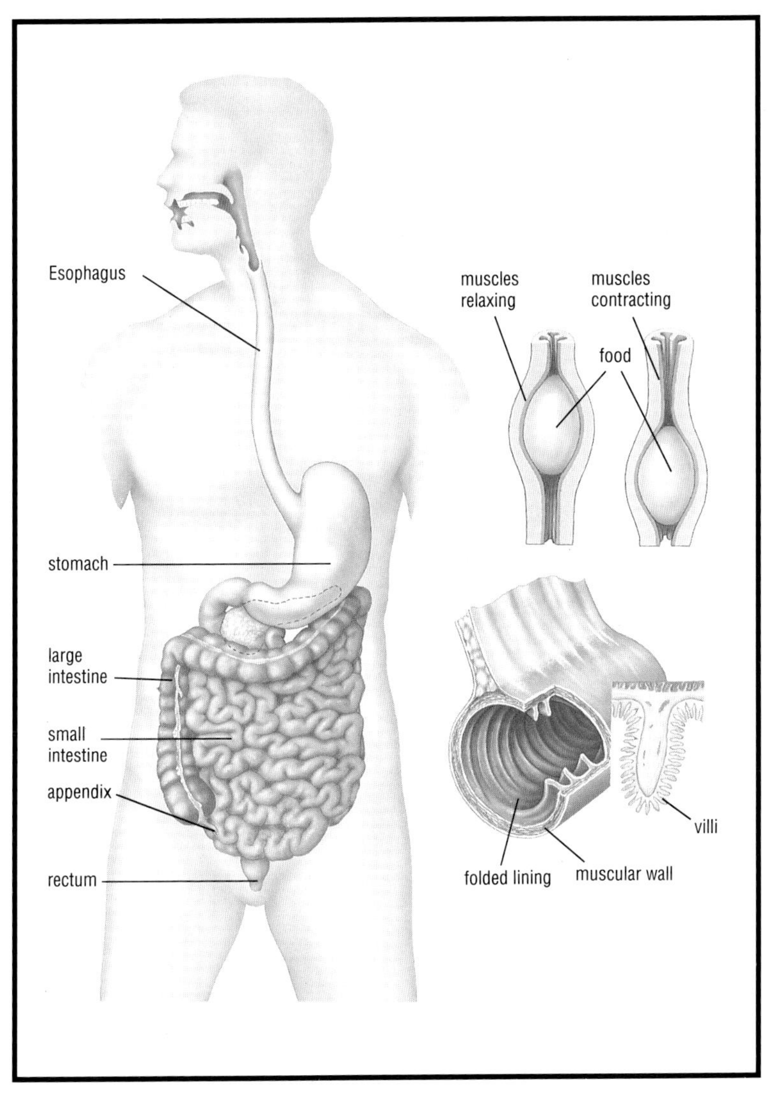

Human digestive system.

Food moving down the Esophagus. Muscles squeeze together behind the food and relax in front, so that food is moved along. Food moves at a speed of 8 inches per second.

The small intestine is lined with a carpet of finger-shaped villi, which give it an enormous surface area for absorbing digested food.

even if you are standing on your head. The food does not fall under gravity but is pushed down by a series of squeezes.

Stomach: Here food is mixed with hydrochloric acid and enzymes. It is squeezed and kneaded for up to 6 hours for a large meal. The stomach can hold about 1 ½ litres (2 ½ pints).

Small intestine: Food is slowly squashed and squeezed along the intestine at about an inch a minute. The tube is 7 m long and about 3 cm (1 ½ in) in diameter. The wall is not smooth but highly folded giving it a much larger area for absorbing food. The surface area is about 9 sq m (100 sq ft), almost five times the area of the skin covering the whole body.

Large intestine (colon): Here water is absorbed from the food remains. This part of the gut is 2 m (6 ½ ft) long and 6·3 cm (2 ½ in) wide.

Rectum: Here the feces collect and are pushed out about once a day. About two-thirds of feces is water, and about half of the rest is bacteria, mainly dead. The bacteria line the intestine and play an important part in the digestion of food.

LIVER

The liver is the largest organ in the body, weighing about 1·5 kg (3·3 lb) in an adult. When we are resting about a quarter of the body's blood supply can be found in the liver.

The main functions of the liver:

● Helping in the digestion of food by producing bile which flows into the small intestine and breaks down fats.

● Storing sugars and starches as a starch-like chemical called glycogen, to be used whenever extra energy is needed.

● Purifying blood by destroying the poisons in it. Removing and storing the iron from dead red blood cells.

KIDNEYS

Our bodies produce many substances that would quickly kill us if we did not have an efficient way of getting rid of them. It is our kidneys that filter out all these unwanted substances and pass them out in our urine.

Each kidney is about 11·5 cm (4 ½ in) long, 6 cm (2 ½ in) wide and 3·5 cm (1 ½ in) thick and only weighs 140 g (5 oz). It contains over a million tiny filter units called tubules.

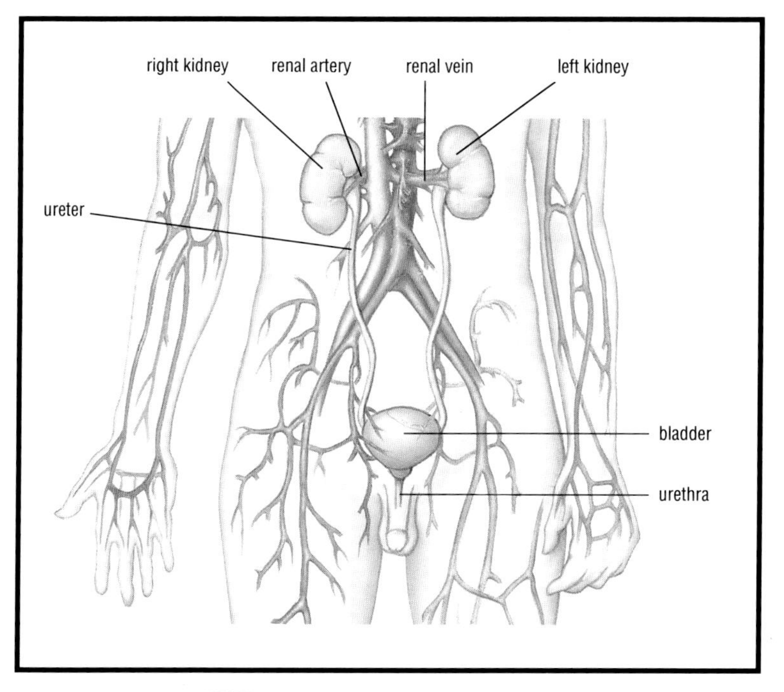

The human excretory system

DID YOU KNOW?

If the filtering tubules in the kidney were laid end to end they would measure more than 300 km (190 miles).

The tubules filter water, food and waste from the blood. They then return to the blood everything that our body still needs. All our blood passes through our kidneys every five minutes. During the course of a day the tubules remove over 200 litres (50 gal) of liquid. Over 99% of it is returned to the blood and just 1 ½ litres (2 ½ pints) passes out as urine carrying all the waste substances.

BLADDER

As the bladder fills with urine it expands into a balloon shape. When it is about 8 cm (3 in) across it contains about ¼ litre (½ pint) of urine and we feel the need to empty it.

THE BLOOD SYSTEM

● Blood is the 'river of life' carrying through your body a wide range of different chemicals. It takes food and oxygen to all organs and transports the poisons and waste products they produce to the kidneys and the liver.

● Blood transports hormones (the body's chemical messengers) from glands where they are produced, to different parts of your body. Some control how much you grow, others will make the body grow a beard or breasts. And when you have to act quickly, perhaps to escape danger, the hormone adrenalin is released into the blood and carried round the body. Adrenalin makes you breathe faster and increases your heart rate, sending the blood racing around the body carrying extra oxygen to the muscles.

● An adult has about 5 litres (8 pints) of blood. About 60% of the blood is a yellow liquid called plasma; it is made up of water containing dissolved food substances, hormones and other important chemicals. Blood has a slightly salty taste because of the salt dissolved in the plasma.

● If you look at a drop of blood under the microscope you will see that it contains lots of tiny red discs. These are red blood cells and it is these that make our blood look red. Their red color is produced by hemoglobin which is a protein containing iron. Hemoglobin combines with oxygen in the lungs, and is transported round the body by red blood cells. There are about 200 million red blood cells in a drop of blood and over 25 million million (25 000 000 000 000) in total in an adult.
Red blood cells only live for about 100 days. New red blood cells are made in the bone marrow which fills the center of larger bones. About 100 million new red blood cells are made every minute.

● Blood also carries the antibodies which fight germs. Floating in blood are white blood cells, which are about four times bigger than red blood cells. There are about 5.5 billion white blood cells in the body. They are constantly on the lookout for harmful germs which they destroy by wrapping themselves round them. White blood cells live only for a few days because they are poisoned by the bacteria they capture. In an infected wound the yellow pus is dead white blood cells that have been poisoned by the bacteria causing the infection.

● Blood travels through the body along an intricate network of blood vessels. The total length of these vessels in an adult is about 160,000 km (100,000 miles). Laid end to end they would go round the world four times, or stretch half way to the moon.

● Blood travels down tiny tubes called capillaries, which are so thin that red blood cells can only just squeeze through. White blood cells and the liquid plasma are actually squeezed out of these tiny tubes and carry food and other chemicals to all the tissues that the blood does not reach. This liquid which washes over the tissues and feeds them is called lymph.

● There are four basic blood groups: A, B, AB and O. Worldwide, group O is the most common (46%), but in some countries, for example Norway, there are many more people with group A. The rarest blood group is AB; only about 14% of the world's population is blood group AB.

Human blood magnified 420 times. The two cells in the center are white blood cells, the rest are red blood cells.

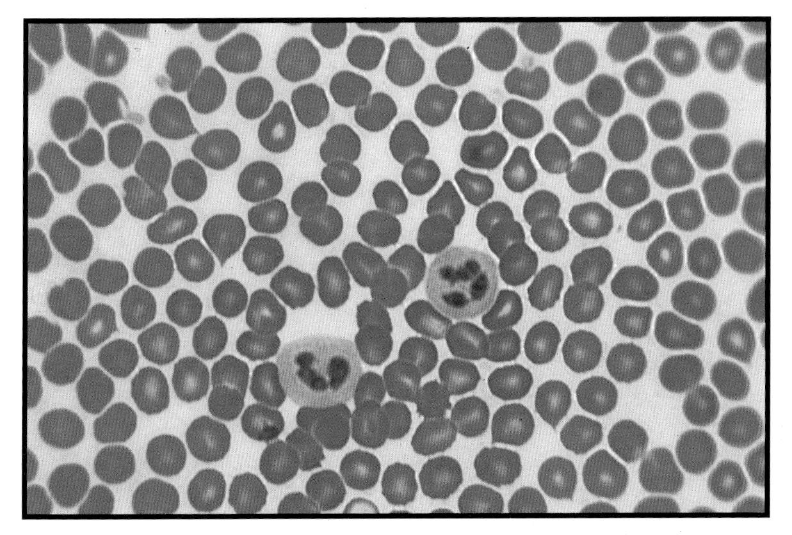

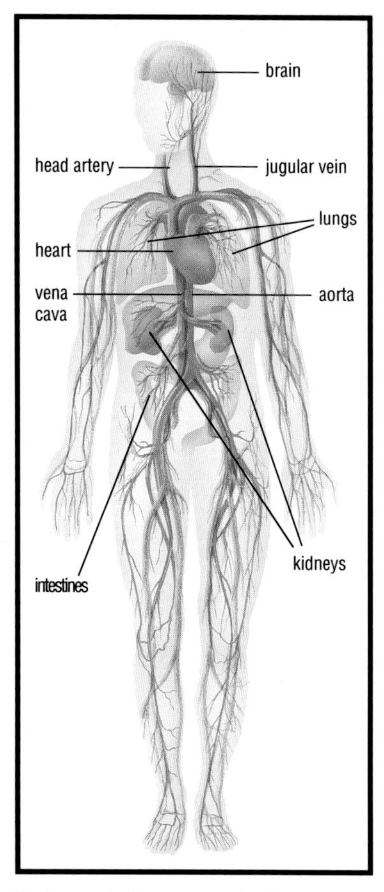

The human circulatory system showing the main blood vessels of the body.

HEART

● Your heart is an amazing pumping system which began its work when your were 25 days old in your mother's womb, and measured little more than half a centimetre (a quarter of an inch) long. Your heart will continue beating until the end of your life, usually without any repair or maintenance. On average it will have pumped more than 200 million litres (about 352 million pints) of blood round your body.

● Every day your heart pumps about 43,000 litres (75,680 pints) of blood, enough to completely fill over 150 baths. The weight of the blood pumped in a day is 43 tonnes (43 tons). This is almost equal to seven times the weight of a fully grown male elephant.

● As a newborn baby your heart races at about 120 beats a minute. This decreases until as an adult it beats at about 70 times a minute.

DID YOU KNOW?

In an average lifetime the heart will beat over 2½ billion times.

● During exercise the number of heart beats increases to two or three times its normal value and the heart also increases the amount of blood pumped at each beat. Normally the heart pumps all the body's blood through it in one minute. During exercise the blood passes through the heart five times every minute. The blood is racing round taking oxygen from the lungs to the muscles. But during vigorous exercise the muscles are greedy for oxygen and the blood supply to some parts of the body, for example your digestive system, almost closes down completely. Some muscles get about 20 times the amount of blood they normally receive when you are resting.

● Your heart is really quite small for all the work it does. In an adult it is no bigger than a clenched fist and weighs 300 g (about 10 oz). When you were born your heart only weighed 28 g (1 oz).

● Your heart is really two pumps side by side. One sends blood to your lungs to pick up oxygen; the other sends this oxygenated blood round the rest of your body.

● Your heart is not on the left as most people think. It is in the middle of your chest with about one-third on the righthand side.

The inside of the heart. The walls of the heart are muscles which contract to force blood through. The valves ensure that blood flow is one-way only.

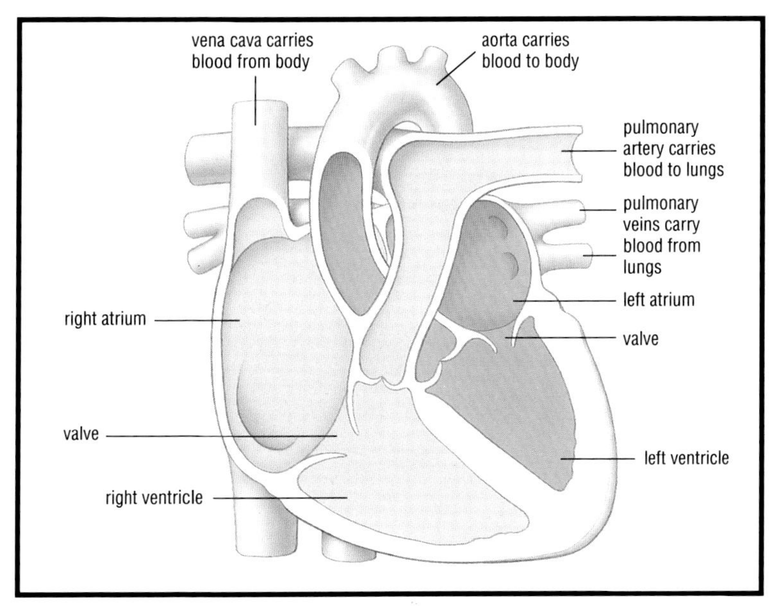

BREATHING

● You can live for a few days without food but without air you will die in a few minutes. It is the oxygen in the air which keeps you alive. When you breathe in air goes into your lungs. The oxygen from the air passes into your bloodstream and is carried by the blood to all parts of your body, where it combines with sugars to produce energy. Carbon dioxide and water are the waste products from this process. They are taken by the blood back to the lungs and breathed out.

● A pair of lungs weigh just over 1 kg (2·2 1b). The right lung is slightly larger and heavier than the left lung. Lungs do not have any muscles. We breathe using the muscles between our ribs and a sheet of muscle called the diaphragm. When breathing in the diaphragm moves down and the ribs move up and out so that the size of the chest increases. The pressure in the lungs is less than outside and air rushes in to equalize the pressure. To breathe out the muscles relax, the chest becomes smaller and air is forced out of the lungs. Fortunately we breathe automatically and do not have to think about it.

● The network of tubes, called bronchioles, in the lungs looks rather like the stalks of a bunch of grapes after the grapes have been removed. At the end of each tiny tube is a small sac called the alveolus. There are hundreds of millions of alveoli in the lungs.
Each alveolus is covered with a tracery of fine capillaries, each capillary only just wide enough for a single red blood cell to pass through. The walls of the alveolus are very thin, less than 0·001 mm (0·00004 in) and this enables oxygen to pass into the blood and become attached to a red blood cell very easily.

● There are several hundred thousand million capillaries carrying blood through the lungs. If laid end to end they would reach 2,400 km (1,500 miles). It takes about one minute for all your blood to pass through the lungs.

● When resting an adult breathes in and out 10 to 14 times a minute. In a day 10,000 litres (2,200 gal) of air pass into the lungs. About 94 gallons of oxygen are removed and 78 gallons of carbon dioxide are

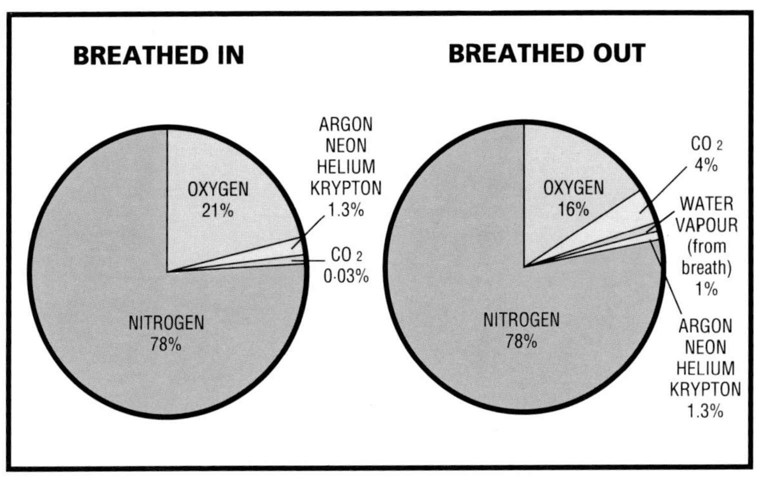

breathed out. For someone working hard those figures are doubled and during vigorous exercise ten times the amount of air is breathed each minute.

Amount of oxygen used by different organs when resting. To help you imagine these quantities, remember that one teaspoon equals 5 millilitres.

Organ	Millilitres per minute
Liver	67
Brain	47
Heart	17
Kidney	26
Muscles	45
Other	48

The gas composition of the air you breathe in, compared with the air you breathe out. Only 5% of the oxygen you breathe in is actually used.

The brain uses the same amount of oxygen whether we are thinking hard or just resting. Our muscles may need 100 times more oxygen when exercising than when resting.

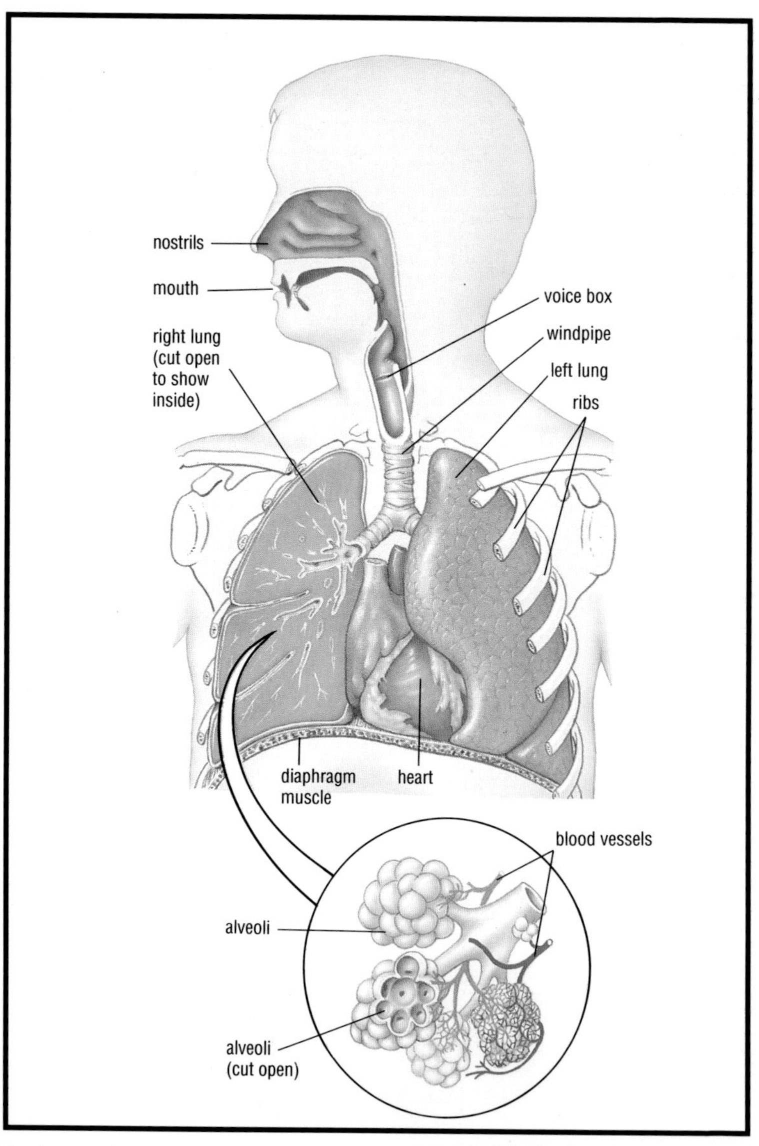

Your lungs are in your thorax. The thorax is the part of your body from your neck to your diaphragm. The lungs are made up of millions of tiny tubes leading to air spaces called alveoli. The alveoli are covered with blood vessels.

BONES

● Without bones you would slump to the floor like jelly. But your bones are much more than an elaborate coat-hanger round which your body is draped. Bone is living tissue with nerves and blood vessels running through it. The bone provides a reservoir of important minerals such as calcium and phosphorus which are needed for many of the body's chemical processes. Bones are changing all the time with minerals flowing in and out of them.

● The number of bones in an adult is about 206 but it does vary; for example about one person in twenty has 13 pairs of ribs instead of the normal twelve. Babies are born with about 305 bones and as they grow up 99 are lost as they join with others.

● The longest and strongest bone in the body is the thigh bone (femur). Its length is about a quarter of your height. It can take a load along the bone of over a ton, but a sideways blow can break it much more easily.

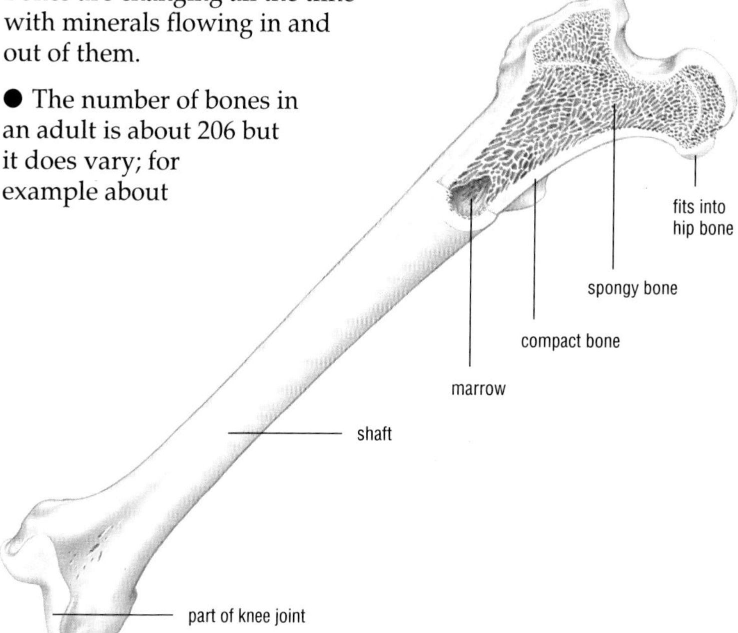

The thigh bone, the longest bone in the body.

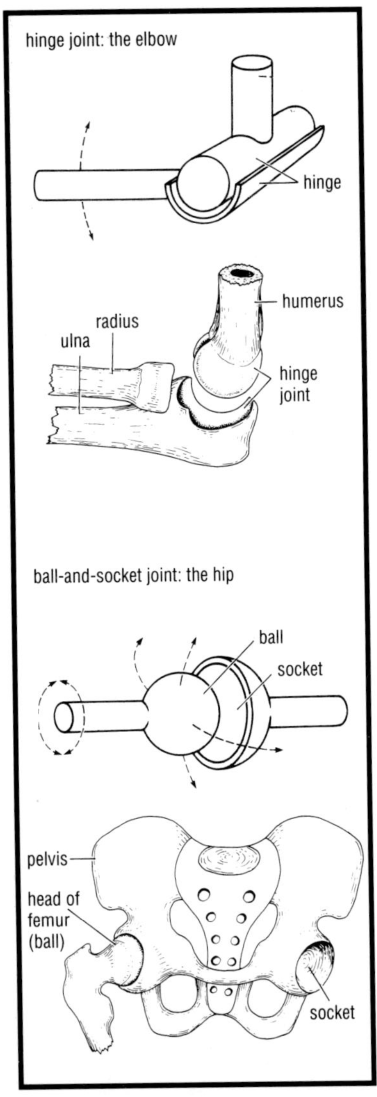

Joints are the bending and sliding places where bones touch. Different types of joint allow different types of movement. Hinge joints allow movement in two directions. Ball-and-socket joints allow movement in many directions.

- The skeleton of a newborn baby is made mainly of a tough slightly flexible substance called cartilage. As it absorbs calcium and phosophorus compounds the bone becomes hard. If you used an acid to dissolve these compounds out of our bones you would be left with a rubbery structure. Some cartilage never changes to bone: the end of the nose and the outer part of the ear, are examples.

DID YOU KNOW ?

The smallest bone in the body is the stirrup bone (stapes), one of three little bones inside the ear. It is about 3 mm (0·14 in) long and weighs about 3 mg.

- Bones are held together by strong bands called ligaments and joined by muscles which make bones move. Where two bones meet they are separated by a pad of cartilage which is lubricated with a slippery liquid that allows bones to move smoothly.

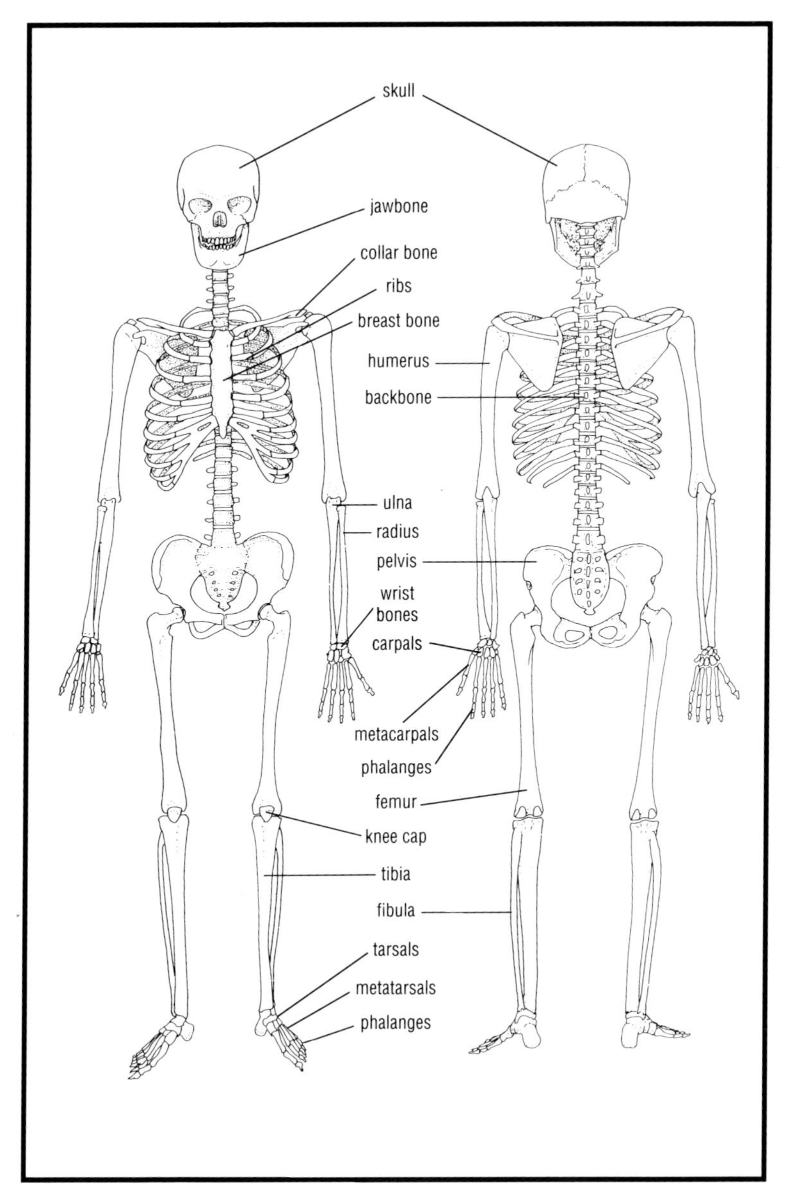

Front and back views of the human skeleton.

MUSCLES

- There are three types of muscle in the body: skeletal or striped muscle which we use to move; smooth muscle which is found in our guts and squeezes the food along; and heart muscle which is different from all other muscles in the body. The heart muscle works non-stop for the whole of your life without becoming tired.
- 40% of body weight is muscle.

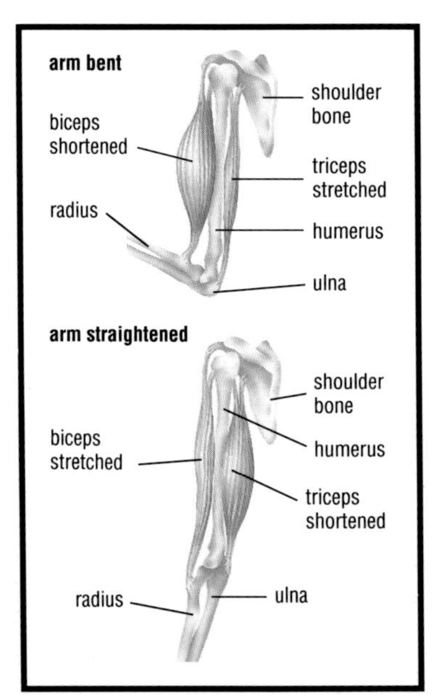

Muscles can pull but they cannot push. So to move your forearm you need two separate muscles: the biceps to pull the forearm up, and the triceps to pull it down.

- Striped muscles are the main flesh of the body, and the meat people eat from other animals. They are called voluntary muscles because we can control them by thinking about them. Most people cannot control the involuntary, smooth muscle or heart muscle. But some experts at yoga can speed up and slow down their heart rate.
- There are 639 named muscles in the body. About 600 of these are voluntary and so under our control.
- The largest muscle is the *gluteus maximus* which extends from the buttock down the thigh.
- The smallest muscle is the *stapedius* which controls the tiny stapes bone in the ear. The muscle is less than 1·3 mm (a twentieth of an inch) long.
- When walking you use 200 different muscles.

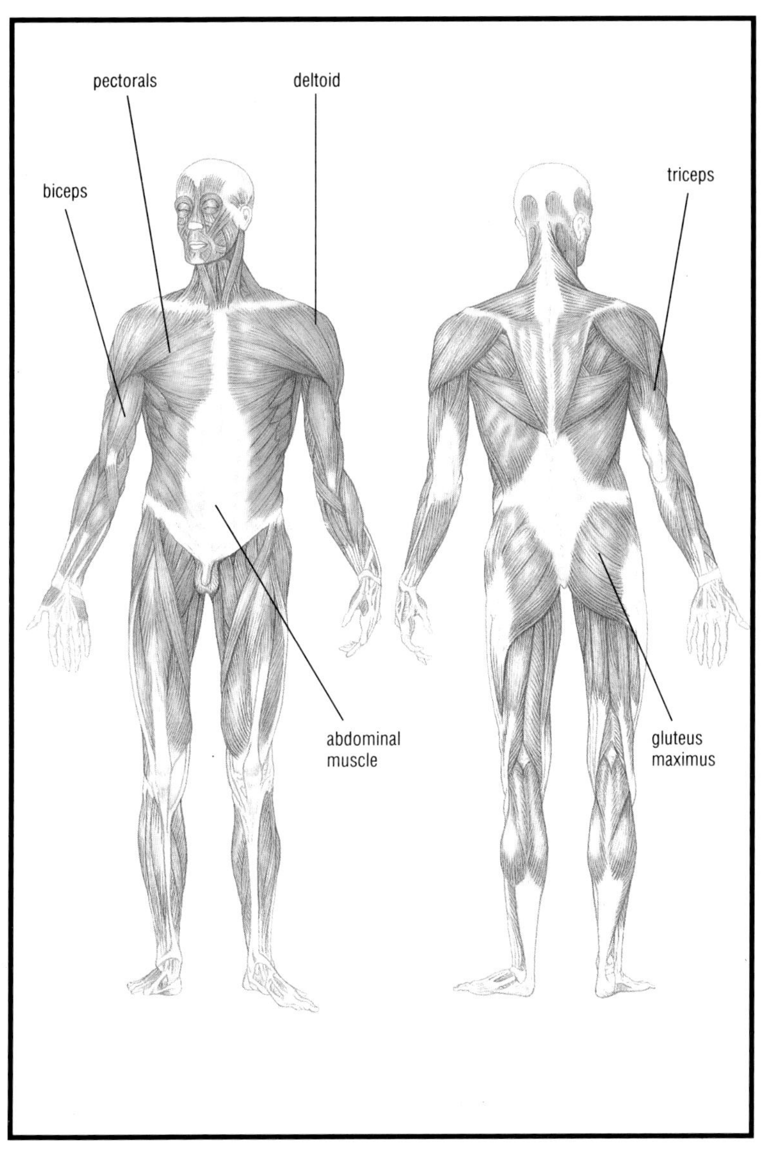

The muscles of the human body, front and back. There are more layers of muscle beneath some of those shown.

THE BRAIN

● The brain is the central control system of the body. It receives and sends messages to organs and muscles along a network of nerve fibers. The messages are sent as tiny electrical impulses which can whizz along at speeds of up to 100 metres per second (300 ft/sec).

● Nerves consist of a bundle of nerve fibers each of which belongs to a single nerve cell, called a neuron.

A nerve is made up of a bundle of nerve fibres. The nerve fibres carry messages in the form of electrical impulses and so are insulated by a fatty sheath. Nerve cells are connected end to end across gaps called synapses.

● Each neuron has a cell body containing the nucleus, and a long fiber, called the axon, which carries electric impulses from one neuron to another. Axons can be tiny but the longest runs from the spinal cord down to your toes and can be over 1 m (3·3 ft) long. It takes less than a fiftieth of a second for an itch signal to travel from your toe to your brain. The brain then transmits an enormous number of complicated signals to bring your hand into position to give a scratch.

● There are two types of nerves: sensory nerves carry messages to and from the skin and other sense organs; motor

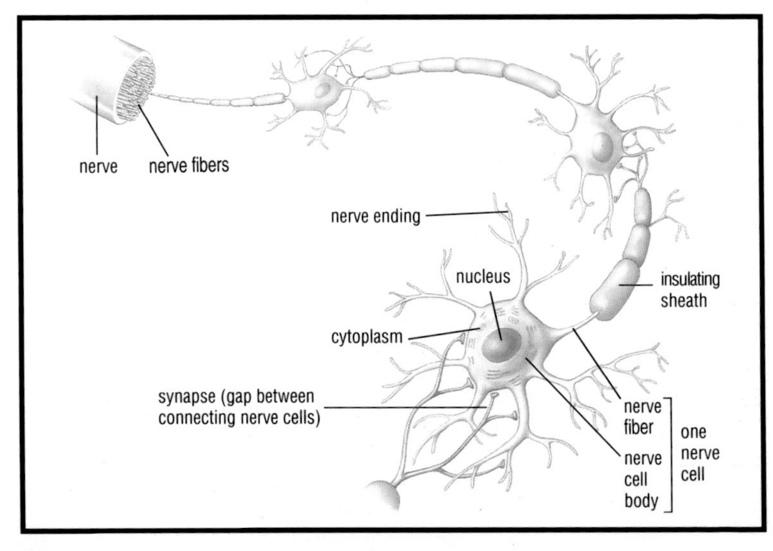

nerves communicate with the muscles making them expand and contract so that we can move.

- There are gaps, called synapses, between nerve cells. Here the electric impulse is converted into a chemical which passes across the gap and triggers an electrical impulse in the next nerve cell.

- The brain has nerves which directly connect it to the organs of the head but most nerves are connected to the brain via the spinal cord. But the spinal cord is much more than just a high speed route for electrical impulses to and from the brain. The spinal cord can control some of our actions without the brain being used at all. These actions are called reflex actions. When we touch a hot saucepan a message is sent along a sensory nerve to the brain. The brain sends a message along the motor nerves to the muscles which pull the hand away. If the saucepan is very hot the spinal cord will send a message to the muscles and the hand will move away before the brain has even received the original message about the hot saucepan.

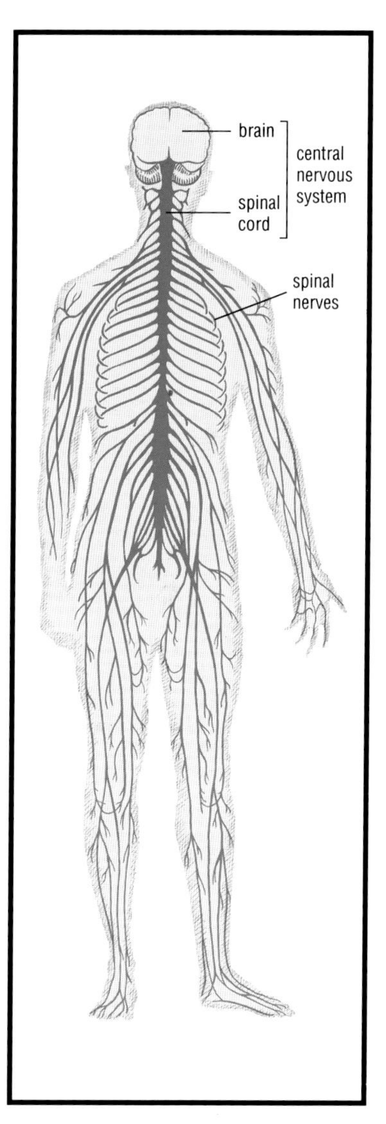

Humans and other vertebrates have a central nervous system made up of a brain and a spinal cord, which are surrounded by bone. Nerves run out from the central nervous system to all parts of the body.

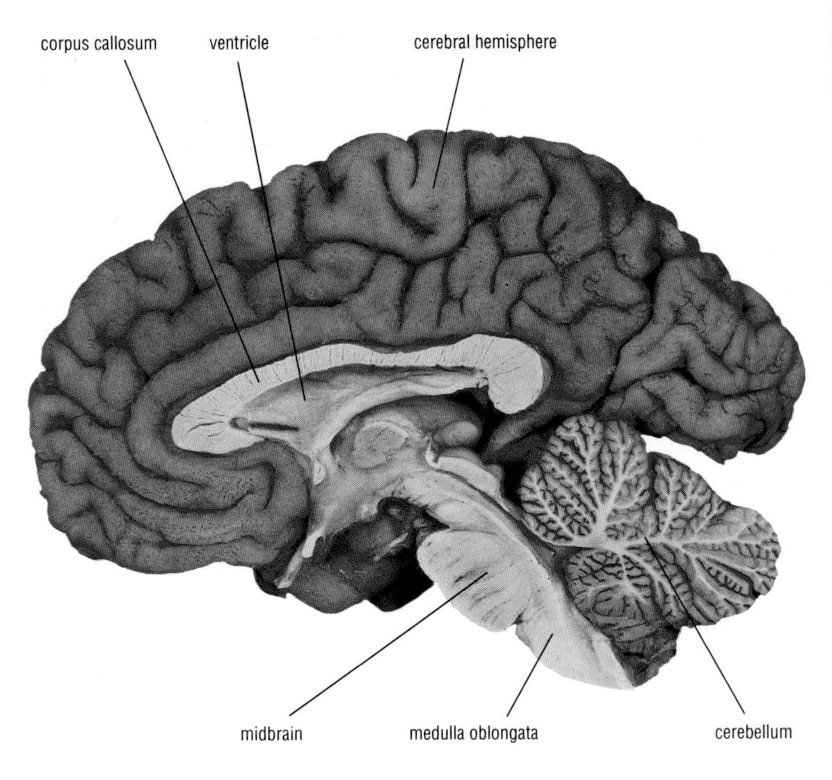

The cerebral hemispheres are involved in thought, memory and the senses. The cerebellum is involved in balance and muscle co-ordination and the medulla oblongata controls breathing and blood pressure.

● The size of an electrical impulse sent along a nerve is always the same. How then does the brain know when something is very hot, or very painful? The only thing that can be altered is the number of signals sent: there can be as many as several hundred, or sometimes even a thousand, in a second.

THE DIFFERENT PARTS OF THE BRAIN

● The cortex is the wrinkled layer of 'grey matter' on the outer surface of the cerebrum. If unfolded it would cover an area of 75,000 sq cm (80 sq ft). It is 3 mm (0·14 in) thick and is where all the data from the senses are processed. It is also involved with intelligence and learning. Different areas of the cortex are responsible for different activities.

● The two hemispheres (or halves) of the brain are not exactly identical in structure, or in the work they do. The right side of the brain seems to have more influence over artistic, musical and creative activities. The left is better at problem solving and dealing with numbers and words.

● The brain needs a great deal of energy to do its work. Although it only weighs 2·5 % of the whole body, it uses about one fifth of all the energy produced.

The brain receives about 9 gallons of blood every hour (over a pint a minute).

● Without oxygen brain cells are quickly damaged and are unable to repair themselves. After about five minutes without oxygen the whole brain is dead.

● As humans get older thousands of brain cells die every day and are not replaced. Fortunately we have thousands of millions of brain cells.

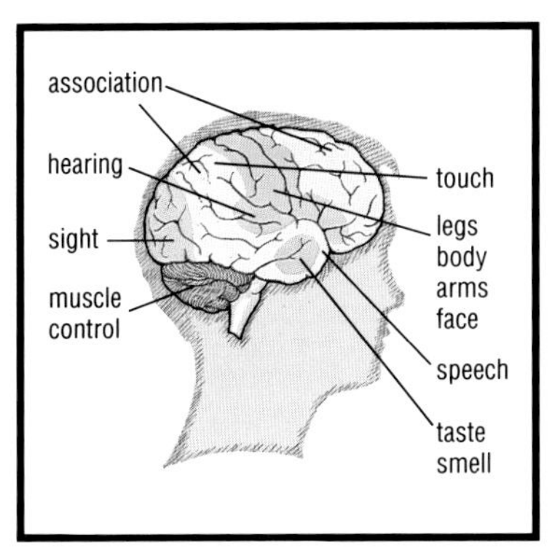

Different parts of the body and different functions are controlled by special areas of the brain. The association areas are where information from the senses and from past experience are put together.

● The weight of an adult brain is about 1·4 kg (3 lb). An average male brain has a volume of about 1·5 litres (about 2 ½ pints). Women's brains are on average smaller but this does not mean women are less intelligent than men!

● There is a machine, called an electroencephalograph (EEG for short) which can measure the electrical activity in the brain. It shows that the brain produces rhythmical changes in electrical voltages known as brain waves. There are four main types of wave:
alpha waves at 10–12 cycles per second occur when relaxed but not during sleep;
beta waves at 15–60 cycles per second appear when we are awake and concentrating;

theta waves at 5–8 per second are produced when we are drowsy;
delta waves at 1–5 cycles per second are large waves produced during sleep in adults but all the time in young babies, even when they are awake.

● By studying EEG patterns of the brain's electrical activity while we are asleep, it has been discovered that we go through different phases of sleep every 90 minutes. Most important are the periods when we dream and our brains burst into activity and our bodies twitch. Our eyeballs show rapid eye movement (REM) at this time and so this is known as REM sleep. If prevented from having REM sleep we soon become irritable and unable to think properly. Most periods of dreaming sleep last about 10 minutes, but just before waking we may dream for as long as 45 minutes. If woken during the dream most of us will remember the dream. Although there have been reports that claim we can learn during our sleep by listening to tape recordings, this is not true. But it does seem that we remember things very well when we think about them just before going to sleep.

An EEG machine measures brain activity and produces print-outs like this, which show how different brain wave patterns are produced by different activities.

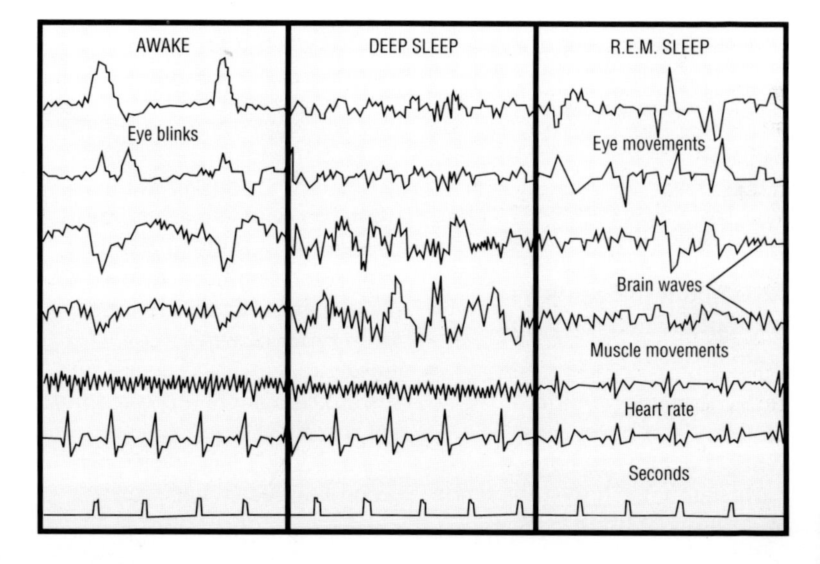

SKIN

● We are all encased in a protective armor of hardened dead cells. In fact when you look at someone all that you can see of them is dead. Their hair, skin, nails, even the surface of their eyes, are all made of dead cells.
The outer layer of dead skin is called the epidermis. Underneath is the dermis which contains the sweat glands, and little pockets, or follicles, from which hair grows. It also contains all the different types of nerve endings and all the tiny blood vessels which keep the dermis alive.

● Skin cells live only for about three weeks.

● The dead skin on our body surface is constantly being shed as flakes. Tens of thousands of flakes fall from us every hour. These tiny, very light particles form most of the dust in your home. They take hours to drift slowly to the floor where they become food for a fungus which is devoured by the microscopic dust-mite.

● The weight of skin covering an adult body is about 3 kg (about 6 ½ lb).

A small piece of skin (magnified). The epidermis, the upper dead layer, wears away every time you touch anything. But the dermis, the layer of live dividing cells, replaces it as fast as it is removed. The dermis also repairs cuts and other damage.

● The thickness of skin covering most of your body is 2 mm (0·1 in). It is thinnest on your eyelids where it measures 0·1 mm (0·05 in) and is thickest on the palms of your feet and hands where it is up to 3 mm (0·14 in) thick.

● The area of skin on an average man is nearly 1·8 sq m (11 ½ sq ft).

● The color of your skin depends on a brown pigment produced in the skin. This pigment, called melanin, is formed by skin cells called melanocytes. Everybody has about the same number of melanocytes but those of dark-skinned people produce more melanin than those of light-skinned people.
When the skin is exposed to strong sunlight more melanin is produced making light skin tan.

Some people have areas where melanin concentrates, producing freckles which may increase in number and darken in the sun.
You may notice dark blotches on the back of the hands of old people. They appear because the melanocytes are no longer producing melanin evenly over the whole surface of the skin.

● When you cut yourself the skin is torn and blood leaks out from broken blood vessels. Immediately platelets (bits of bone marrow cells floating in the blood) stick together and collect at the edge of severed blood vessels. These platelets,

When you bleed, platelets send out fibers which trap red cells. Blood then changes into a thick jelly, a blood clot, which blocks the wound.

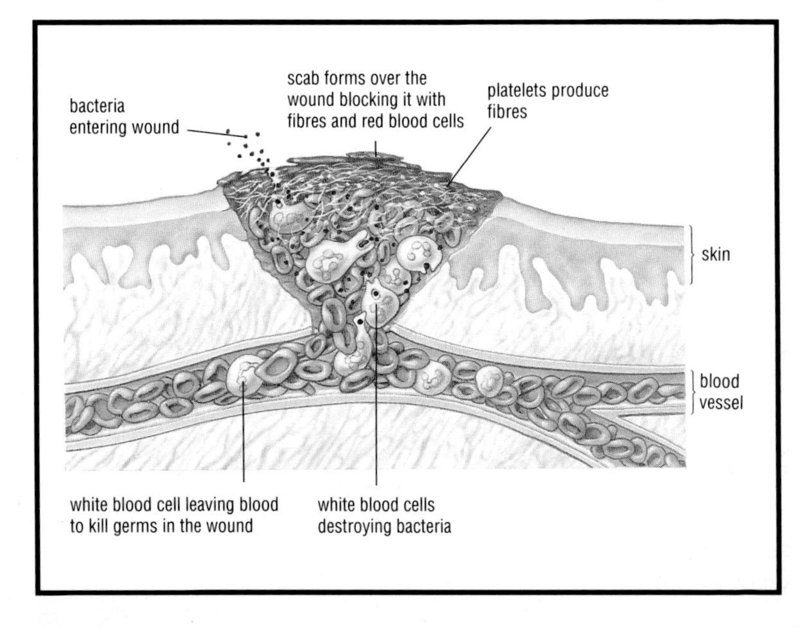

only about 2-4µm in diameter, block the open end of the blood vessel. At the site of the cut a web of thin threads of a substance called fibrin, traps blood cells. When dry it forms a protective scab beneath which new skin grows.

HAIR

● Each hair grows from its own little pocket, called a follicle, in the skin. Only the bottom part of the hair in the follicle is alive. New cells are produced here and push up the old dead ones. Attached to each follicle is a tiny muscle which pulls the hair upright when you get cold, and gives your skin 'goose pimples'. In animals with fur these muscles have an important function because the fur traps more air close to the skin when the hairs are upright, and that helps keep the animal warmer. We do not have enough hairs to get any benefit from the movement produced by these muscles.

● The shape of the follicle determines the type of hair you have. Wavy hair grows from oval-shaped follicles; straight hair from round ones; and curly hair from flat ones.

● Our first follicles develop when we are 7 week old babies in the womb. The first follicles appear on the eyebrows, chin and upper lip. At 16 weeks the head is covered with the beginnings of hair.

● A baby in the womb grows hair all over its body. This soft, silky hair does not have any color and is usually shed before birth.

● Babies have many more hair follicles on their head than at any other time in their lives. At birth there are about 1200 per sq cm (nearly 8,000 per sq inch) this decreases to 800 per sq cm by the age of one.

DID YOU KNOW?

An average scalp has 100,000 hairs, and about 60 fall out every day.

● Hair on the scalp grows at about 0·35 mm per day or 12 cm (4 ¾ in) a year. Hair can grow up to 90 cm (3 ft) before it falls out. The follicle then rests for about 4 months before a new hair begins to grow.

● Boys' hair grows faster than girls' hair, and boys have more hairs per square inch on the scalp than girls.

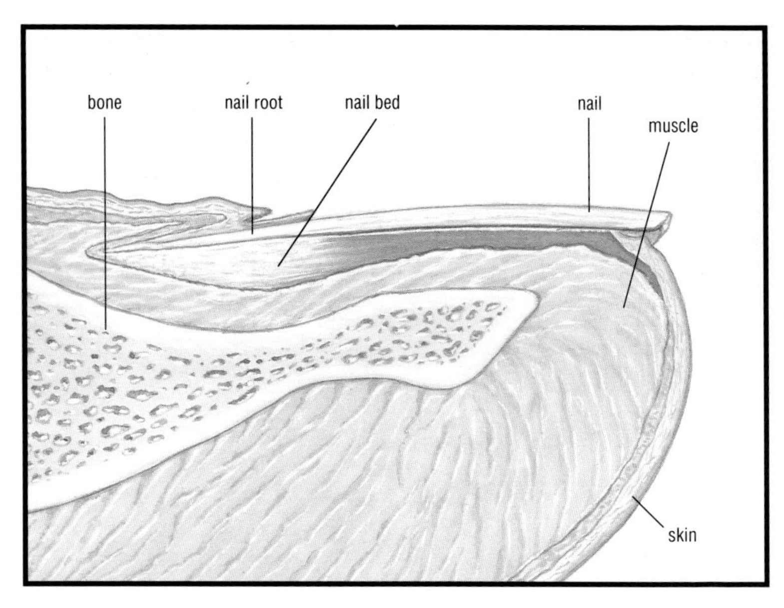

Your nails have three parts: the root, the bed, and the part of the nail you actually see. The root produces the cells for the nail and the bed, so it is where growth takes place.

● Hair is quite stretchy and can be pulled gently to almost twice its original length.

NAILS

● Nails are made of keratin, the same substance found in hair, horns, hooves and feathers.

● Fingernails grow about 0·5 mm (0·02 in) a week. If never cut or broken, fingernails would reach about 2 m (over 6 ft) in a lifetime.

● In right-handed people fingernails grow faster on the right hand than on the left. (This study did not include any left-handed people). They also grow faster on men than women and grow more during the day than night.

● Nails on different fingers grow at different rates. The nail on the middle finger grows fastest; the slowest are those on the thumb and little finger.

● Toenails grow at about 0·1 mm (0·004 in) a week. If never cut or broken during your lifetime, they would reach about a third of a metre (about 12 in).

TOUCH

● The sense of touch really includes all the sensations we feel with our skin: touch, pressure, pain, heat and cold.

● The way we feel things in our skin is not well understood. There seem to be special nerve endings in the skin: some can detect warmth, others cold; some are sensitive to touch, others to pressure and pain. But it is not always that simple because we can feel all these things on our ears where there are none of these nerve endings.

● Certain parts of our body are very sensitive to touch and they have large concentrations of touch receptors, the nerves which receive messages when touched, and send them to the brain. Our backs are rather insensitive and only have receptors every 63 mm (2 ½ in). On our fingertips there is a receptor every 2·5 mm (0·1 in) and on the tip of the tongue they are 0·6 mm (0·025 in) apart. Any small object in the mouth can easily be detected by the tongue, and a new tooth-filling can feel enormous.

● The skin can detect tiny vibrating movements. It is most sensitive to vibrations of 200-400 cycles per second. The frequency of a tuning fork is in this range.

● There are astonishing accounts of people working with machines and having an arm sliced off without feeling any pain at the time, or even being aware of what has happened. It seems that if you are concentrating on some activity you are not sensitive to pain.

● In an experiment patients in a hospital suffering severe pain after an operation were given tablets that contained only sugar and no pain-relieving medicine. But the patients were told that they were getting medicine and one-third of them no longer experienced any pain, just as if they had really been given medicine.

● People who have had an arm or a leg amputated still feel pain in their toes and fingers even though they no longer have any. This is called phantom pain. When the limb is removed the nerves are cut through but continue to send messages to the brain. The brain is tricked and makes the person feel as if they still have the complete painful limb.

SENSES

SIGHT

● An eyeball weighs only 7 g (¼ oz) and its diameter is 2·5 cm (1 in). Male eyes are slightly bigger than female eyes by about 0·5 mm (0·02 in).

● The pupil at the center of the eye opens in the dark to let extra light into the eye. Pupils also increase in size for other reasons, including fear and interest. It has been measured that a man's pupils increase by 30% at the sight of a beautiful woman!

● Our eyes adapt amazingly well in the dark. They become about 75,000 times more sensitive to light.

● The lens in the eye focuses a tiny upside-down image of the object on the retina at the back of the eye.
In an eye test you look at letters on a chart at a distance of 6 m (20 ft). The smallest letters on the chart are 8·75 mm (0·35 in) high. The image they make on the retina is 0·025 mm (a thousandth of an inch) high. With good vision your brain

When you look at an object, the light from it passes through the lens and is focussed on the retina. Nerve cells in the retina pass messages to the brain. The image of the object is formed on the retina upside down, but the brain interprets the messages the right way up.

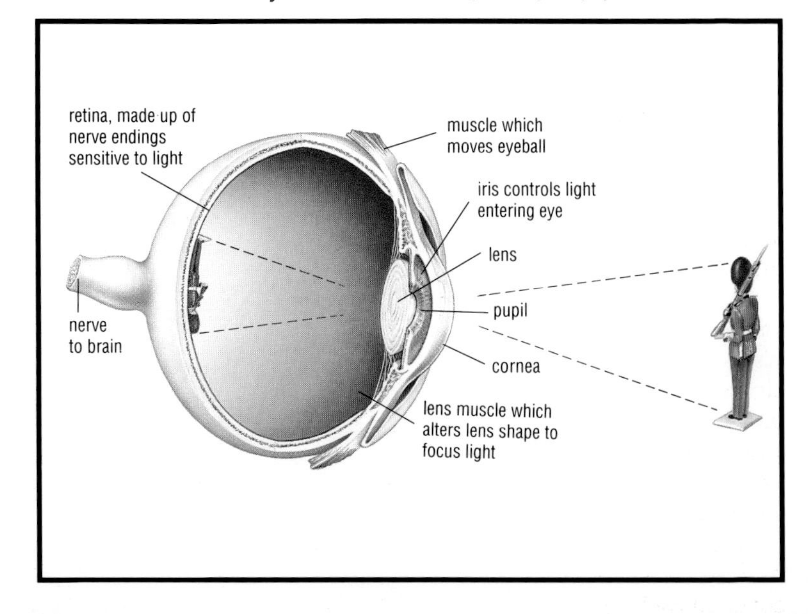

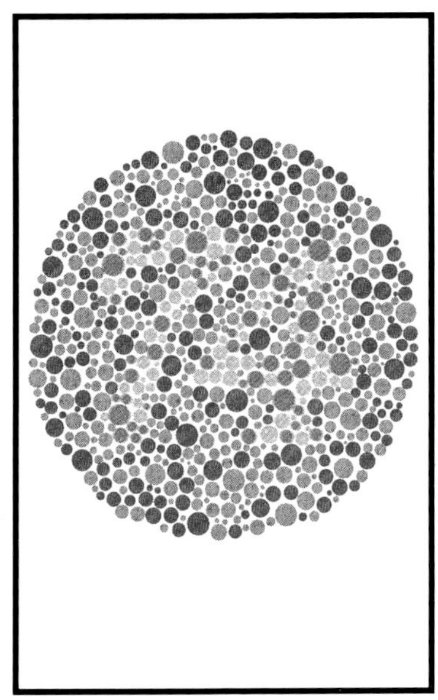

Color blindness chart: those who are unable to distinguish between red and green cannot read the number.

will be able to tell exactly which letter you are looking at.

● On the retina at the back of the eye there are special receptors which receive light and transmit information to the brain. There are two types of receptors: rods and cones. The rods are used mainly for night vision and can only detect shades of grey. The cones are used in daylight and different cones are sensitive to red, green or blue and, in combination, allow us to see different colors. There are 127 million rods and 7 million cones in each eye.

● The sensitivity of the eye is staggering. In good light, using both eyes, we can distinguish up to 10 million different shades. We can also judge the position of an object with great accuracy.

● Color-blindness is found mainly in men. About 8% of European and white North American males are color-blind and less than ½ % of females. Color-blindness in other groups is lower: only 3% of black males and 1% of Inuit males. The commonest form of color-blindness is where red and green are not easily distinguished.

● It is impossible to stop yourself from blinking except for a very short period of time. Normally a blink takes place on average every 6 seconds and the eyes stay closed for about a sixth of a second. This means your eyes are closed for about 2 minutes in every hour you are awake. In a year you will have blinked about 8 million times and blinking will have kept your eyes closed for over 170 hours. Blinking helps keep the surface of the eye clean and moist.

● When very young babies stare at you it is noticeable that

they do not seem to blink. Babies do not begin to blink regularly until they are about six months old.

● To keep our eyes moist we produce about 200 g (a teacupful) of tears in a year. Tears contain an antiseptic which helps protect our eyes from bacterial infection. All mammals produce tears to cleanse their eyes. But humans seem to be the only animal in the world that cry for emotional reasons.
Babies do not produce tears when they cry until they are several weeks old.

Sounds make the eardrums vibrate, which makes the ossicles vibrate, which makes sensory hairs in the cochlea vibrate, which sends nerve impulses to the brain where we hear sound.

HEARING

The outer ear ends in a very thin membrane called the eardrum. When sound waves travel down the ear canal and hit the eardrum they make it move. In order to hear a sound the eardrum only has to move 0·000 000 01 mm (40 billionths of an inch).
The movement of the eardrum is passed onto three tiny bones which transmit the vibration to another membrane called the oval window. These vibrations are passed into the cochlea, which is a coiled tube filled with liquid and lined with tiny hair-like sense cells. As these move they send electrical signals to the brain where the jumble of signals are analyzed

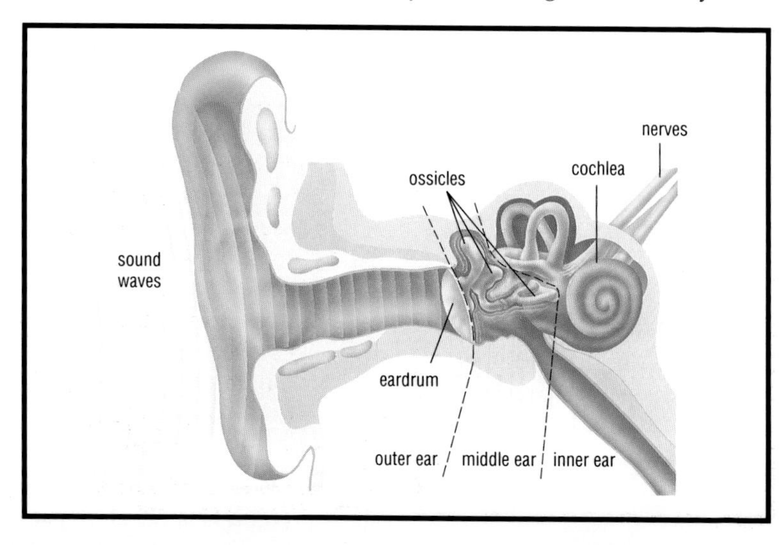

and interpreted by the brain as a sound.

● One of the most important functions of the ear is controlling our sense of balance, so that we do not topple over. The semi-circular canals near the cochlea are filled with liquid and when this liquid moves messages are sent to the brain, which then makes sure we keep upright and balanced. But you can trick your brain by spinning round fast. When you stop, the liquid in the semi-circular canals continues moving. Your brain thinks you are still moving and everything you look at carries on spinning so that you feel dizzy.

● We are very good at sorting out the direction of a sound. With two ears working together we find the direction by noticing the difference in the timing of the arrival of the sound at each ear. We can detect differences as small as 0·0001 second (one ten-thousandth of a second) and use this information to help locate sound.

The loudness of a sound is measured in decibels (dB). Very loud sounds can damage your hearing. Listening to loud music on headphones, or standing too close to loudspeakers at rock concerts, can cause some deafness.

● The range of sounds we can hear is between frequencies of 20 and 20,000 vibrations per second, this is about 10 octaves. Top C on a piano is 4096 vibrations a second, and 20,000 vibrations a second is heard as a hiss. Above this range is ultrasonic sound that we cannot hear but many other animals can. For example dogs come running when they hear dog whistles at 35,000 vibrations per second; bats can detect frequencies of 75,000 vibrations per second; and bottle-nosed dolphins can identify sounds from 20 to 150,000 vibrations per second. Dolphins do not receive these sounds through their external ears. They 'hear' by detecting vibrations through their jaws.

SMELL

● Your nose is very sensitive. High up inside the nose is the olfactory organ which is responsible for your sense of smell. It consists of two areas of cells which have tiny hair-like detectors attached to them, called cilia. The cilia wave around in a layer of thick liquid called mucus. All molecules of smell have to dissolve in the mucus before they can be detected by the cilia and a message is sent to the brain.

● Sniffing is necessary for air to reach the olfactory organ. During a sniff air is inhaled at 12 km per hour (20 mph).

● We are able to smell the difference between more than 10,000 different odors.

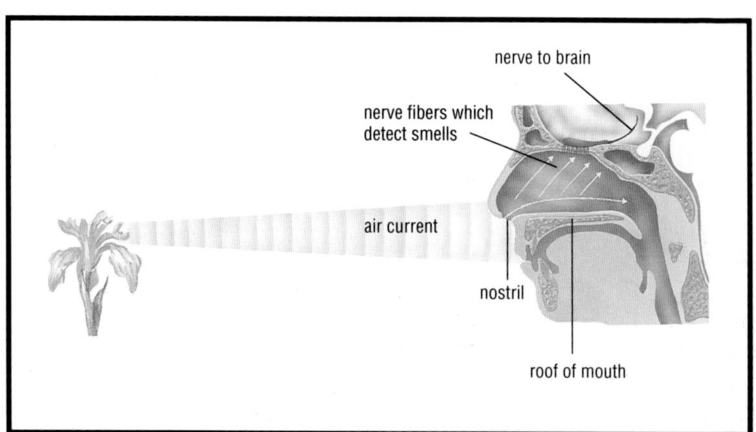

Chemicals in the air you breathe dissolve in the moisture-covered nerve ends in your nose. These chemicals make the nerves send messages to your brain which produces your sense of smell.

● Babies seem to have an especially sensitive sense of smell, and at 6 days old can smell the difference between milk from their mother and any other milk.

● Mothers can recognize their own baby's odor.

● Our sense of smell may be impressive but that of a dog is a million times better. If you took 1 g (one thirtieth of an ounce) of a smelly substance in our sweat called butyric acid and mixed it with all the air in a ten-story office building you would just about smell it. If you took that gram of butyric acid and mixed it with all the air in a 100 m (300 ft) layer above a large city, a dog would be able to detect it.

● We all produce our own particular odor. This is shown most clearly when tracker dogs follow the scent of one person. Highly-trained dogs can even distinguish between identical twins.

● Do you know that you can act rather like a tracker dog, and follow the smelly trail left by feet? Put an absorbent paper such as blotting paper, on a firm floor and get someone to walk over it barefoot. Now get down on your hands and knees and by smelling you should be able to track them!

● Experiments with college students have shown that they can recognise the odors of other students in their class, and members of their own family.

● Scientists have shown that we can smell whether men or women have been sitting on chairs after they have left the room. It has been suggested that our sense of smell might, without us realizing it, affect who we choose as friends.

TASTE

● There are four basic sensations of taste: sweet, bitter, sour and salt. All flavors are a combination of these four.

DID YOU KNOW?

You have about 3,000 taste buds in your mouth. They are not only found on the tongue but also on the roof, cheeks and back of the mouth.

● Our sense of taste is not nearly as sensitive as our sense of smell. Sensitivity is

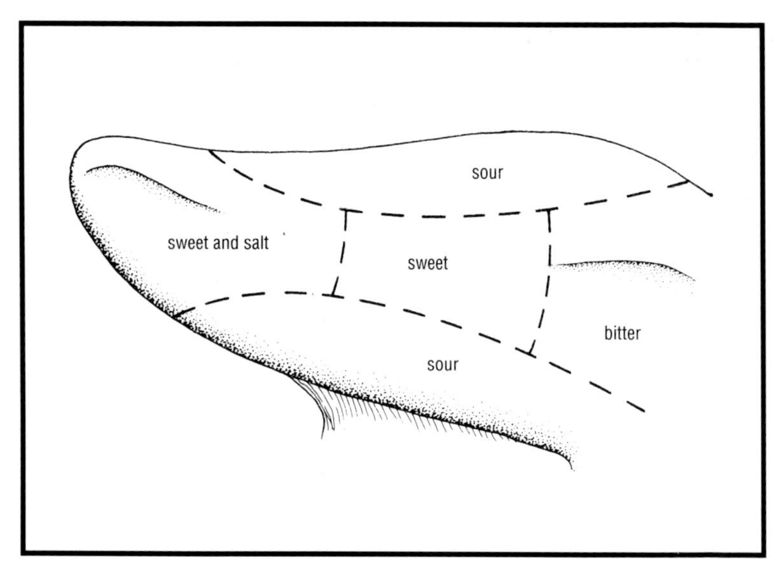

The tongue is covered with taste buds containing nerve endings which transmit taste information to the brain. The tongue can only distinguish between four basic tastes.

measured by finding out how much we can dilute a substance and still taste it.

● We can detect the four basic substances in the given dilutions:

Taste	Dilution
bitterness	one part in two million
sourness	one part in 130,000
saltiness	one in 400
sweetness	one in 200

We are therefore most sensitive to bitterness and least sensitive to sweetness. But compare these figures with the fact that we can smell some substances at one part in more than a billion.

● For detailed information about the food we are eating we rely on our sense of smell. When we have a blocked nose, as with a cold, we find our food tasteless.

● Experiments have shown that babies in the womb like the taste of sweetness. Newborn babies give a slight smile when given something sweet. If a bitter substance is tasted they make a face.

THE LIFE THAT LIVES ON US

● A close inspection of your body with a microscope will reveal all sorts of creatures growing and feeding on you: bacteria, fungi, mites, perhaps a flea from your cat, and maybe even some lice in your hair.
Most of them are very small and are measured in units called micrometres (μm).

Here is a millimetre. ▪ 1mm

There are 1000 μm in a millimetre.

Sometimes the number of organisms growing in a square centimetre is given.

Here is a square centimetre. (1 cm × 1 cm)

FUNGI

● The most common type of fungus found on your body is a yeast called *Pityrosporum ovale.* It is oval in shape and measures about 2 μm wide and 4 μm long. It lives happily on your hair and greasy areas of your skin such as the scalp and around the nose. As with all yeasts it is a single-celled fungus that reproduces by growing a daughter cell which breaks off. It is fortunate that this fungus is completely harmless as you can often have as many as half a million in a square centimetre.

● The fungus called thrush, *Candida albicans*, can move into damp areas of the body, causing itching and unpleasant smells. Babies sometimes have it on their bottoms and even as white patches on their mouths. It is easily treated.

● Fungi thrive in moist places such as hot sweaty feet where the fungal growth is called Athlete's foot; and around the groin where men sometimes get jockstrap itch.

● Blood and tears both contain a chemical which prevents fungal growth.

BACTERIA

● Even when clean your skin is swarming with thousands of millions of bacteria. Depending on the type of bacteria they may measure between 1 μm and 4 μm.

Bacteria on the surface skin of an adult man:

Part of body	Number of bacteria in a square centimetre
scalp	1,500,000
armpit	1,000,000
forehead	200,000
back	50,000
forearm	11,000

● Inside the follicles and oil glands at the base of the hair are bacteria which do not need oxygen to live. There can be ten times as many of these as the ones living on the skin's surface.

● In ideal conditions bacteria breed at an amazing rate. One bacterium can divide in two in twenty minutes. So in three hours there could be 500 bacteria and in 8 hours about a million.

● Moisture helps bacteria breed. In a few days a bandage on your arm will increase the number of bacteria in a square centimetre on your skin from a few thousand, to tens of millions. Now you know why you must keep cuts and grazes dry and preferably uncovered.

● Washing with soap and water removes some bacteria but most are left on the skin. But do not jump into a bath of disinfectant to get clean. The bacteria on your body are harmless and actually prevent other nasty bacteria from moving in and taking over.

● Certain bacteria produce unpleasant smells or body odor (B.O.) when they feed on the sweat and oils which ooze onto our skin. Most anti-perspirants contain aluminium compounds which block up the glands which produce these secretions. The bacteria are killed by many deodorants but this can allow fungi and nasty bacteria to move in.

● We are constantly launching large colonies of bacteria into the air. It has been estimated that a milligram of skin scales have over half a million bacteria; and you lose a gram of skin a day.

DID YOU KNOW ?

In 24 hours, a billion skin scales fall from your body carrying several hundred million bacteria.

MITES

● Tucked away in the follicles, (small pits), from which your hair grows are tiny, eight-legged follicle mites called *Demodex folliculorum*. The body of this mite is about a third of a millimetre (.012 inches) long and bends in the middle so that the mite can climb in and squeeze itself between the hair and the wall of the follicle.

● Almost everyone has mites in the follicles of the nose, the scalp and eyelashes. Some eyelashes have been pulled out and found to have 25 mites gripping to them with their tiny clawed feet.

● One of the most troublesome mites to humans is the dust mite *Dermatophagoides pteromissimus*. It is not found on our bodies but feeds on the skin scales we shed continually. These skin scales make up 90% of household dust. Almost half of all people with allergies are allergic to the dust mite, or rather what it excretes.

FLEAS

● Fleas are hardly a problem to us in our warm, dry and clean homes. But cat and dog flea pupae are found lurking in the corners of the carpet that the vacuum cleaner cannot reach. Given a chance, animal fleas will happily make a home on humans, and feed on their blood. Fleas can easily hop from one animal to another, and one species of flea makes jumps of up to 8 inches. This distance is 150 times the flea's own length.

A cat flea feeding on a human. The female flea must have a blood meal before she can lay fertile eggs.

LICE

● The head louse *Pediculus humanus capitis* is 1/8–3/16 inches long, and feeds on your blood every three hours. We do not feel the tiny jabs it makes in the scalp as it pierces the skin. But the louse injects its saliva into the open wound, probably to stop the blood clotting; this saliva sets up an irritation which in about a week makes us want to scratch.

Human body lice. The louse on the left is swollen with blood after feeding; the other is about to feed. Body lice look similar to head lice but are slightly larger and now much rarer.

● The adult female head louse lays about 10 eggs a night for up to 28 days. These whitish eggs, or nits, are nearly 1/16 inch long and are firmly stuck to the hair near the scalp. In girls nits are usually found just behind or above the ears. In boys they are found near the top of the head.

● Adult lice on blonde hair are paler than those on dark hair.

● Lice cannot be washed off hair. In water the louse closes the breathing holes it has on either side of its body. It survives long periods of holding its breath.

BED BUGS

● These brown, blood-sucking insects are about 1/4 inch long. They can live in mattress seams and in cracks in walls and furniture. At night they creep out and feed on blood sucked from the sleeper's face or throat. Fortunately they can be destroyed by insecticides and do not live in clean houses.

Bed bug feeding on a human. If temperatures are high, around 77°, bed bugs must feed every night. If the temperature is very low the bed bug can go for 500 days without a meal.

ANIMALS

Mammals

Reptiles

Amphibians

Birds

Fish and other sea creatures

Insects and other arthropods

MAMMALS

● All mammals feed their babies on milk produced by the female in her mammary glands. In the platypus these glands ooze milk onto the skin and the babies lick it off. In kangaroos and most other marsupials the milk comes from teats in the mother's pouch. In sheep, goats and cows the milk is contained in the udder near the back legs. In cats, pigs and most other mammals there are nipples along the underside of the animal. In humans the female's breasts are the mammary glands.

● All mammals are warm-blooded and all of them have some fur or hair on their bodies. Sea mammals, such as whales and dolphins, have a thick layer of fat, or blubber, to stop them getting cold.

● All mammals have lungs and breathe air.

● All mammals have bones including a backbone.

● When we talk about animals most of us think first of mammals. But mammals make up only a tiny number, just 0·3%, of all animal species. Most animals are insects!

EGG-LAYING MAMMALS

● A small number of mammals lay leathery-shelled eggs. These are the monotremes, found only in Australia and New Guinea. There are only three species of egg-laying mammal: the platypus and two species of echidna.
The **platypus** is an unusual mammal: it is poisonous and it lays eggs. When the young hatch the mother does provide them with milk but not from nipples. She has glands in the skin of her belly, rather like sweat glands, which ooze milk, and the young lick it off her fur.
The male platypus has a poisonous sting. The poison is released from a spur behind the ankle of each back leg, so the platypus can give a poisonous kick to any predator chasing it in water. A scratch from the spur can kill a dog and be really painful for a human.

● **Echidnas** are spiny anteaters with powerful claws for tearing apart the nests of ants and termites. If frightened they will use their claws to dig a hole straight down into the ground and within a few seconds only sharp spines can be seen.

When the female starts to lay her single egg she curls her body tightly into a ball so that she can deposit the egg into a temporary pouch on her belly. The egg is damp and sticks to her fur. In about a week it hatches and feeds on the milk that oozes from its mother's skin. The young remains in the pouch until its spines begin to grow, but it will continue to lick milk from her for some weeks.

DID YOU KNOW?

The adult platypus does not have teeth, but its babies grow three tiny teeth which they keep for only a short time.

MAMMALS WITH POUCHES

● Opossums and kangaroos both give birth to babies that look little more than blobs of jelly. They have no fur, their eyes and ears are tight shut and only their front limbs are well-developed. A female **kangaroo** weighing 32 kg (70 lb) gives birth to a baby only about the size of your fingernail. It crawls up through its mother's fur into her pouch; a journey that

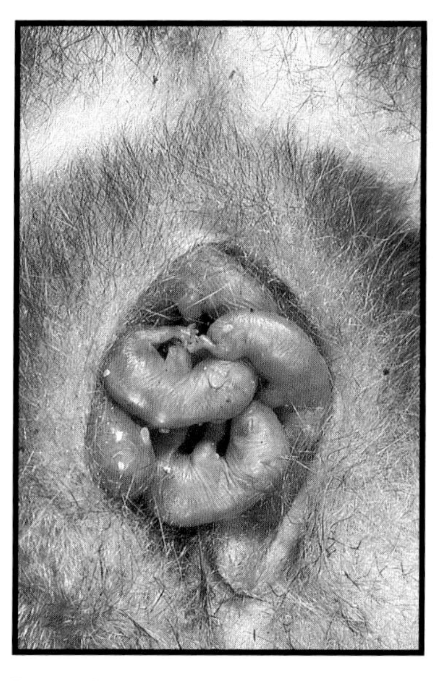

Baby southern opossums attached to teats within the pouch. Southern opossums are found in Central and Southern America.

takes about three minutes. Once in the pouch it fixes on to one of her four teats. The teat swells in the mouth of the young kangaroo, so it cannot fall off, and is supplied with milk.

The female American **opossum** gives birth to more than twenty babies. They are so tiny that all together they hardly fill a teaspoon. They clamber the 8 cm (3 in) journey to their mother's pouch. Half of them do not make it. The Virginia opossum has thirteen teats in her pouch and any baby that does not get a teat starves and dies. About 10 weeks later the babies climb out and are carried on their mother's back.

● A baby **koala** lives in its mother's pouch for about 6 months, and then spends the next 6 months carried on her back.

● The Australian **marsupial mole** has a pouch for carrying her babies. It opens backwards so that it does not fill with earth as she burrows underground.

The pygmy shrew is one of the smallest mammals. Shrews are very active and consume large amounts of food for their size.

INSECT EATERS

● Some types of shrew have poisonous saliva so that a single bite can kill or paralyze their prey. Most shrews eat insects and worms but the American short-tailed shrew can overcome quite a large frog with its venomous bite.

● The **pygmy shrew**, found in parts of Europe, is one of the smallest mammals. It is no more than 6 cm (2 ½ in) long, and so thin that it can crawl down tunnels about the width of a pencil.

● Shrews eat all the time and will generally eat anything

The North American star-nosed mole has a nose which is divided into 22 moveable, fleshy tentacles that are highly sensitive.

they can get their teeth into. When insects are in short supply they will eat other shrews, or even their own young. If they do not have food for just a few hours they die, so they can only sleep for short periods.

● Shrews communicate with each other using high-pitched squeaks. They also produce sounds in the ultrasonic range, well above frequencies that we can hear. These noises help them find their way around, like the method of echo-location used by bats.

● **Moles** eat large quantities of insects and worms. They can eat their own weight of earthworms a day. If a mole finds extra worms in its underground tunnels, it bites the back of them; this does not kill the worm but stops it wriggling away. The mole keeps these live, disabled worms in a larder tunnel so that it always has a supply of fresh worms. Stores have been uncovered with thousands of worms in them.

● Moles are not blind, but their eyes probably only allow them to tell the difference between light and dark.

● The **star-nosed mole** of North America has an elaborate arrangement of feelers round its nose. They may also provide the mole with an extra-sensitive sense of smell.

● **Hedgehogs** hunt at night using their keen sense of smell to find insects, snails, and small animals such as frogs. They also have a very good sense of hearing.

TOOTHLESS MAMMALS

● The slowest mammal is the **two-toed sloth** of South America. It is only active for about four hours a day and spends almost all its life hanging upside down in the trees moving very slowly, covering about 1 m (3 ft) in 15 seconds. It climbs down to the ground once a week to defecate, and produces exceptionally foul-smelling excrement.
The sloth never grooms itself, so that groups of parasitic moths live undisturbed in the depths of its thick coat. They produce caterpillars which feed on the sloth's hair.
The sloth's fur is brown or grey, but appears green because of the small algae plants growing in it. This helps camouflage the sloth among the trees. Sloths seem to have very poor hearing. If a gun is let off right next to it, the sloth does no more than slowly move its head round and blink.

● Although the **giant anteater** of South America is toothless, its powerful hooked claws have a savage grip. There is a story that a battle between a jaguar and an anteater left them both dead. The jaguar's teeth had gashed the anteater, which still held the jaguar in its vice like grip.

● The giant anteater has a tongue 60 cm (2 ft) long, that can collect 500 ants or termites with one lick. It can eat several thousand at each meal.

BATS

● Bats are found all over the world, except the coldest parts. About 70% of all bats eat insects.

● There are bats which feed on nectar and play an essential part in pollinating many tropical plants. In the wild banana trees are only pollinated by bats.

● **Vampire bats**, found in central and South America, are blood drinkers. Some types feed on birds, others on cattle,

There are 951 species of bat. This small selection gives some idea of their variety.

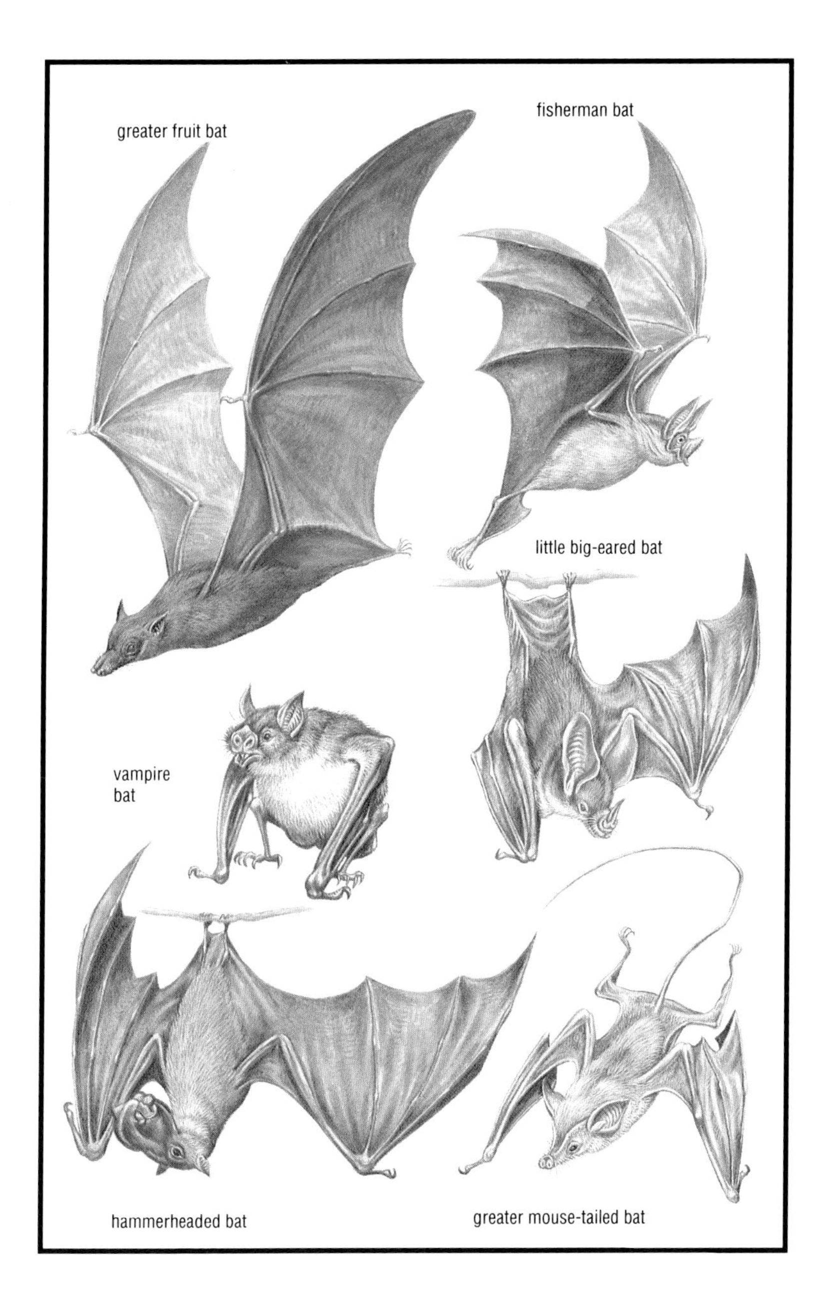
greater fruit bat
fisherman bat
little big-eared bat
vampire
bat
hammerheaded bat
greater mouse-tailed bat

but none feed regularly on humans. They never grow to more than 9·5 cm (3 ¾ in) but are dangerous to animals because they can carry rabies, and can spread this killer disease. With a single bite, they inject the animal with a chemical that stops the blood from clotting, so that the blood continues dripping and they can feed.

● Bats can emit and detect very high-pitched sounds. These ultrasonic noises bounce off surrounding objects and the time taken for the return of the echo gives the bat precise information about its surroundings. One bat has been shown to be able to detect the thickness of a human hair, using this method of sonar echo-location. Different bats produce and detect frequencies from 20,000 up to 130,000 vibrations per second.

● **Fish-eating bats** can detect a tiny ripple produced by a fish. They swoop down plucking the fish from the water with their sharp claws.

● The largest colony of bats in the world is found in the Bracken Cave in Texas. Between 20 and 40 million Mexican **free tail bats** roost there every day. At night they leave the cave and fly round the countryside, feeding on 250 tons of insects each evening. Each bat can devour up to 1000 mosquito-sized insects in an hour.
In the summer they breed. A female gives birth to a single baby, or pup. All the pups hang together on the ceiling of the cave forming a giant nursery with up to 300 tiny, pink hairless pups covering each square foot. The cave is over 180 m (600 ft) long and yet each mother finds her own baby among this seething mass of pups, and goes to it several times a day to suckle it.

● **Flying-fox bats** do not use sonar to find their way around. They find over-ripe fruit to feed on using their good eyesight and highly-developed sense of smell. The largest have a wingspan of over1·7 m (5 ½ ft).

DID YOU KNOW ?

One of the smallest mammals is Kitti's hog-nosed bat, found in Thailand. It has a wingspan of 16 cm (6 in) and weighs less than 2 g (0·07 oz).

East African crested porcupine. To warn off intruders the porcupine shakes its short tail, which is tipped with hollow quills, and makes a loud rattling noise.

PLANT EATERS: RODENTS

● Rodents have gnawing teeth that never stop growing. They include animals like mice, rats, beavers and porcupines. By gnawing at plants they wear down the teeth and keep the edges very sharp.

● African **porcupines** are very gentle creatures that only eat the woody parts of plants. They have been seen gnawing bones, but it is thought they do this only to sharpen their teeth, or provide some phosphates for their diet. Its quills protect the porcupine from enemies. Quills easily break off once implanted in an attacker's flesh and the wound often becomes infected.

● The largest rodent is the **capybara.** Fully grown it can weigh about 60 kg (130 lb) and measure 1·3 m (50 in) in length. Capybaras live in South America.

● **Hamsters** dig underground tunnels, with the main burrow 2–3 m (6–10 ft) below the ground. There are several chambers: one is used as a nursery for the young hamsters; and there is a large chamber used as a larder. More than 30 kg (65 lb) of seed and grain have been found in one of these larders. Hamsters are found wild in the grasslands of central Europe and central Asia.

● The West African **giant rat**, like all plant eaters, needs a

The Giant African pouched rat has cheek pouches similar to those of a hamster. As well as being used for carrying food the cheek pouches may be filled with air as part of a threat display.

great deal of plant food to provide it with enough nutrition. But life outside its burrow is dangerous. To avoid spending too long collecting food this rat rushes out after dark and crams its cheek pouches with anything that looks the least bit like food, then it runs back to the safety of its burrow and sorts it all out, making one pile of food and another of stones, bits of wood and other things that are not edible. It is able to carry about 200 possible bits of food in a single journey.

● The **brown rat** is sometimes called the sewer rat. Rats live in a community and help each other. People have seen them helping another that is carrying something too heavy for it. The cries of a rat in a trap warn other rats to stay away.

Brown rats on a garbage pile. Alert, intelligent and highly social, the rat has been much misunderstood.

It is estimated that there are two or three brown rats for every human being on earth.

Flying squirrel gliding between trees. It steers by moving its limbs and using its tail as a rudder.

Three rats eat as much food as one human, though brown rats weigh only about 700 g (1 ½ lb). It is important to control the rat population, but this is very difficult because they have become resistant to even the most powerful poisons.

● **Flying squirrels** have fur-covered skin joining their front and back legs, which spreads out like a parachute and enables them to glide for distances more than 50 m (165 ft). They are nocturnal creatures and have large eyes to help them see at night.

● **Beavers** make lakes by building dams across streams. The land behind the dam floods and provides the beaver with a good place to build its home, or lodge. The lodge is half-submerged in the lake, with an underwater entrance to prevent predators getting into the beaver's cosy home. The dam and lodge are built with trees that the beaver cuts down using its powerful front teeth. Beavers can fell a tree trunk 12 cm (5 in) across in less than 30 minutes. The beaver also stores extra branches in

Canadian beaver next to its lodge. All family members, except kits (the very young), help to build the lodge and keep it in good order. The adult females do the most work.

the dam as a food supply for winter when the lake freezes over.
Beavers repair their dams if there are any leaks. The level of the lake rises as mud is deposited, so the beaver builds a new dam on top of the old one. Some dams are a thousand years old and up to 1 km (3,300 ft) long. Beavers are found in North America, parts of Europe and Central Asia.

● **Lemmings** are small vole-like animals living in Norway. They breed in enormous numbers, even through the winter. When food is plentiful a real population explosion can follow and this produces a food shortage. The lemmings seem to be seized by panic and thousands upon thousands set off on a frenzied migration in search of new territory. Many die of exhaustion trying to swim across rivers and fjords, or they are eaten by predators attracted because of their numbers. They have even been known to leap off cliffs in their search for open spaces.

ELEPHANTS

● Elephants are the largest land animals. The male African elephant grows up to 3·2 m (10 ½ ft) tall and weighs about 6 tonnes (6 tons). Tusks of 3.35 m (11 ft) in length have been recorded, with one pair weighing 200 kg (440 lb).
An elephant eats up to 150 kg (330 lb) a day of leaves, tree bark, grass and roots. It has just one enormous molar tooth in each half of each jaw, which it uses to crush its vegetable diet. After three or four years the molars become worn down and are replaced by new ones, though in very old elephants this ceases to happen.
The woody food takes a long time to digest. A meal we eat takes about 24 hours to pass through our body; an elephant's meal takes 2 ½ days, spending most of that time in its huge stomach. Half of the food is excreted undigested.

DID YOU KNOW ?

An elephant pregnancy is the longest of any animal. It takes up to 22 months for the calf to develop inside the mother

There are only two species of elephant: African and Asian. The two species differ in a number of ways. African elephants are the larger.

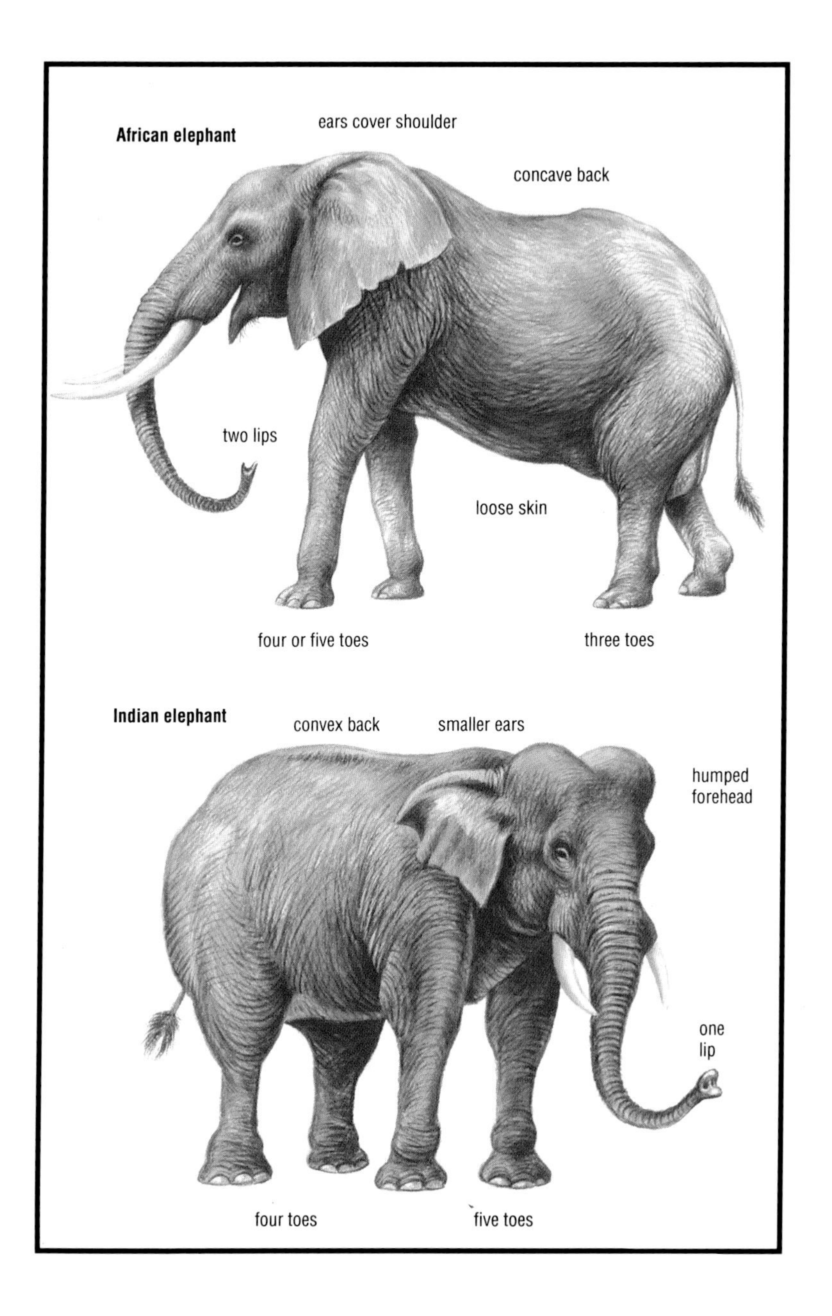
African elephant
ears cover shoulder
concave back
two lips
loose skin
four or five toes
three toes
Indian elephant
convex back
smaller ears
humped
forehead
one
lip
four toes
five toes

Hippopotamus eating grass. If you look closely you can see the pink sweat, that gave rise to the old belief that hippos sweat blood.

● The young elephant feeds on its mother's milk until it is 5 years old.

● Elephants have poor eyesight but a good sense of smell. When you see them swinging their trunks from side to side or lifting them in the air, they are smelling their surroundings.

HIPPOS AND RHINOS

● It is surprising that the **hippopotamus**, with its savage canine teeth eats only grass. The two tusk-like canines can reach 70 cm (28 in) and each weigh 4 kg (8 ½ lb). They are used in fighting with other males. The hippopotamuses feed only at night, spending the day wallowing in water. To prevent their skin drying out when they leave the water they produce sweat which is an oily pink or bright red liquid. Hippopotamuses live in groups of 10 to 150 animals and usually have an adult female as leader.

● The **white rhinoceros** is the heaviest land mammal after the elephant, weighing up to 3·6 tonnes (8000 lb). It grows to 5 m (16 ft) long and its horn can grow to 1·5 m (5 ft) in length. Rhino horn is not made of bone, but very tightly packed hair. Rhinoceroses feed on grass. They have very poor eyesight but excellent hearing and a good sense of smell.

OTHER GRAZERS

● **Camels** survive in the harshest desert climates. They store water in the form of fat in their humps, and do not lose as much water as most other animals when they are hot. The camel produces very little urine and does not sweat until its body temperature reaches 46°C (115°F): a body temperature which would kill us. We keep cool by evaporating water in the form of sweat, but if we lose about one-eighth (12%) of our body weight we die. A camel can lose up to a third of its body weight and just feel very thirsty. There are records of camels carrying heavy loads across a desert without drinking for five days. At the end of the journey one camel drank 110 litres (29 gallons) in one go and another 65 litres (17 gallons) a bit later. The camel visibly swells as it drinks.

● The **giraffe** feeds mainly on the leaves of the thorny acacia tree, eating 14 kg (30 lb) a day. Giraffes spend a great deal of time nibbling. They only sleep for an hour at a time and often do not sleep at all during 24 hours.

The giraffe is the tallest mammal in the world with an adult height of more than 5 m (16 ft). It can weigh more than 1800 kg (4,000 lb) and its long legs can carry it along at an impressive gallop of 47 km/h (29 mph). The female giraffe gives birth while she is standing and the calf drops to the ground head first. It stands up a few minutes later and is running the next day.

MEAT EATERS

CATS

● Cats are superbly designed hunters. They keep their claws sharp by pulling them into protective sheaths. Only when they grab a victim do their claws extend into the animal's flesh. The cat then gives a single bite to the neck of its prey, cutting through the spinal cord and killing it.

● **Jaguars** seem to enjoy going fishing. The jaguar attracts fish to the surface by hitting the water with its tail, and then scoops the fish up and stuns it with its paw.

● The **cheetah** is the fastest land animal reaching speeds of 90–98 km/h (56–63 mph). From a standing start the cheetah can reach this top speed in three seconds, but it is exhausted

lion
snow leopard
tiger
cheetah
jaguar

A few of the 37 species that make up the cat family.

after a fifteen second sprint over a distance of about 400 m (1300 ft).
The cheetah does not roar like the other big cats, but makes a mewing hiss-like sound.

● Small wild cats, including the **lynx** and **ocelot**, purr to show pleasure. The big cats such as lions and tigers are unable to purr. It is not known how cats produce purring sounds.

● **Leopards** are immensely strong and can leap distances of over 6 m (20 ft). When they have killed an animal they carry it up a tree to eat it. Leopards weigh up to 80 kg (176 lb) and have been known to carry animals more than half their weight up trees.

● **Lions** live in groups (prides) of about 6–30 members. There are usually one or two adult males in the group. The males are extremely inactive, resting for about 20 hours a day. They rarely hunt, and do not even eat regularly. Sometimes they go without food for a week and then gorge themselves on 40 kg (88 lb) of meat. Since they weigh about 190 kg (420 lb) this massive meal is about one-fifth of their body weight. If you weigh about 33 kg (70 lb), a lion-sized feast for you would be about 6 kg (14 lb) of meat. Most of the time the males are

Leopards will often carry prey up into a tree, to eat away from scavengers.

provided with food by the females, who hunt for food and look after the cubs. A lioness cannot run as fast as her prey, so lionesses often hunt in groups. They form a circle round a herd of antelopes or zebras, then some of the lionesses run forward driving the animals towards the others, who will probably succeed in making a kill.
The male's main task is to protect the pride from other intruding lions. Occasionally he will give a mighty roar which can be heard 8 km (5 miles) away, just to warn other lions that he is around.

● The **puma** has the largest range of any big predator. It is found from Alaska to Patagonia; in the heat of the Amazon rainforest and in the freezing temperatures at heights of 4,500 m (15,000 ft) in the Andes mountains. The puma eats mainly deer and will bury any unfinished animal, returning to it the next day.

Polar bear female with 10 month-old cubs. At birth cubs weigh less than 1% of their mothers body weight.

BEARS

● **Polar Bears** are powerful swimmers and they have occasionally been seen swimming 320 km (200 miles) from land. Polar bears have a layer of blubber 10 cm (4 in) thick, to prevent them freezing in temperatures that can go lower than -40°C (-40°F). They have hairy soles on their feet so that they do not slide around on the ice, and this enables them to run at speeds of about 38 km/h (19 mph) when hunting. To survive, a polar bear must eat a seal every five days. They will happily feed on rotting meat and can smell a dead whale up to 32 km (20 miles) away. Their sense of smell is about a hundred times more sensitive than ours.

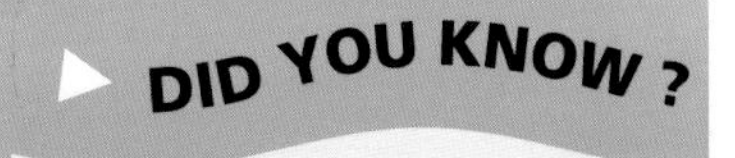

Polar bears are the largest meat-eating animals. An average male is 2·4 m (7 ¾ ft) long and weighs about 400 kg (880 lb).

OTHER MEAT EATERS

● The **honey badger** of Africa has an unusual diet of snakes, turtles and fruit. It also loves honey, and follows the call of the honey-guide bird in order to find wild bee hives, which it then rips open. The honey badger has no predators; its very tough skin protects it from snake bites, and even from the big cats, which kill their prey by biting the spinal cord at the neck. The honey badger's tough coat prevents them getting their teeth into its neck. Also its skin is loose enough for it to be able to turn and bite its attacker even while being held.

A honey badger eating honeycomb, a favorite food.

Spotted hyenas at a kill, Kenya. Ninety per cent of their food comes from animals heavier than themselves.

● **Hyenas** make a mad laughing sound as they prepare to go hunting. They hunt at night in groups of thirty. But sometimes a team of two or three hyenas will work together to kill a wildebeest. They chase after a herd and then stop and watch them, probably choosing a particularly slow or weak animal. Then they chase after that one animal. As one hyena confronts the wildebeest face to face, the other hyena leaps and sinks its teeth into the animal's belly, ripping it open. Their teeth are so strong that they can even eat bones, so the whole wildebeest is devoured. The hyena's powerful jaws can easily break a horse's thigh bone with a single snap. Hyenas live in the dens of warthogs and the burrows of aardvarks, chasing out the original inhabitants and taking over their homes.

● **Wolves** live together in groups, or packs, which consist of two or three couples and their young. The pack is very well organized with one leading male and female. Any member of the pack disagreeing with the leader is quickly thrown out. They hunt as a team and can keep up a steady trot for hours, exhausting the animals they are hunting.

● **Jackals** are found in the grasslands of Africa and central Asia. They hunt in groups but prefer to find dead animals to feed on. They clear up anything left by other meat eaters.

Bottlenose dolphins are found in the coastal waters of most tropical, subtropical and temperate regions.

A few of the 38 whale species (not including dolphins)

SEA MAMMALS

● **Bottlenose dolphins** have large brains and are exceptionally intelligent. They use ultrasound to detect objects around them. They produce sounds with frequencies of about 200,000 vibrations per second, about the same as those produced by bats. These dolphins also make lots of other sounds; about 20 different ones have been identified. Scientists are still trying to discover if dolphins

bowhead whale
white whale (beluga)
blue whale
minke whale
narwhal

can talk to each other in some sort of complex language.

WHALES

- **Humpback whales** are well known for their singing. Each spring they gather in Hawaii to mate, give birth to young whales and to sing. A song can last as long as half an hour and the whales will then repeat it, note for note exactly the same, hour after hour. After a few months in Hawaii the whales swim off to Alaska. The next year they return to Hawaii with a new song. These songs can be heard 30–50 km (20–30 miles) away.

- **Orcas** (also called killer whales), have enormous appetites. The stomach of one 8 m (25 ft) long killer whale had 14 sea-lions inside, and another contained 32 seals! Orcas can grow to over 9 m (30 ft) in length and weigh 4000 kg (8,800 lb). They swim at speeds of up to 20 km/h (12 mph) but can reach 64 km/h (40 mph) for a short spurt.

- The adult female **blue whale** is the largest animal in the world. She can grow to over 30 m (100 ft) long and weigh over 160 tonnes (160 tons). This makes her as heavy as 25 fully grown male African elephants. Blue whales, like most of the larger whales, do not have any teeth. Inside their mouths are strainers called baleen plates, which filter out tiny plankton from mouthfuls of water. Every day the blue whale eats over 4 tonnes (4 tons) of a small shrimp-like creature called krill.

DID YOU KNOW ?

At birth a baby blue whale is 6 m (22 feet) long and weighs 10 tonnes (10 tons), almost twice the weight of an adult African Elephant.

- **Elephant seals** can weigh up to 3,500 kg (7,700 lb) which means that only certain whales, the white rhinoceros and elephants are heavier. They move slowly and clumsily on land but in the sea they can dive to depths of 180 m (600 ft) and stay under water for 12 minutes. They can survive without food for 100 days; imagine not having a meal for three months.

PRIMATES

Primates include the lemurs of Madagascar, the pottos of Africa, the lorises and tarsiers of Asia, the monkeys of South

Male elephant seals fighting during the breeding season. Males can be more than 3 times the weight of females.

America, the monkeys and apes of Africa and Asia and, most widespread of all, human beings.

● The **tarsier** has adhesive pads on its toes so that it can cling to the smoothest branches. It cannot walk, and moves by leaping like a frog. Its huge eyes enable it to see at night when it goes hunting for insects and lizards. Each eye weighs more than the animal's whole brain. In daylight its pupils become so small to shut out most of the light, that the tarsier is almost blind. It can turn its head almost all the way round through 360°.

● The **aye aye** lemur of Madagascar has an extraordinarily long and useful middle finger on its front paws. It uses it to eat with, dipping it into fruits or water and then sucking off the liquid. It also uses it to hook out wood-boring insects from inside the branches of rotting trees.

Aye aye with young. This is a rare sight as the aye aye is now almost extinct.

Bald uakari monkeys live in groups of up to 30 animals, and feed on fruit and leaves.

MONKEYS

● Male **emperor tamarin** monkeys help the females give birth. The male washes the babies and keeps them warm in the fur on his back, giving them back to their mother every two or three hours for feeding.

● **Howler monkeys** have a very large voice box and can make their throat swell and vibrate with sound. A chorus of howlers can be heard several miles away. They probably make the loudest animal noise in the world.

● The **pygmy marmoset**, with a body length of only 10 cm (4 in) and a weight of 50 g (1 ¾ oz), is the smallest monkey in the world.

● The **bald uakari** spends its whole life at the top of the tallest trees in the Amazon rainforest. It has a bright red bald face which becomes paler when the animal is unwell.

● **Japanese macaques** have shown amazing intelligence. Scientists were particularly impressed by one female studied in the wild. She liked to eat sweet potatoes and one day when they were dirty she

The 133 species of monkey are enormously varied. These are just a few.

pygmy marmoset
douroucouli
proboscis monkey
black spider monkey
mandrill
vervet monkey

washed them in the sea. When given rice mixed with sand she threw the mixture into water where the sand sank. She then skimmed the floating rice off the surface of the water and ate it. Her whole group learnt what to do by watching her.

APES

● **Gibbons** have arms as long as their bodies and legs together. They travel through the trees swinging from branch to branch, leaping distances of up to 10 m (33 ft). They live in a family group of parents and up to four youngsters. Every morning the whole family sing together with loud hootings.

● The **orangutan**, found in Borneo and Sumatra, is the heaviest tree-dweller in the world. A male can grow to over 1·5 m (5 ft) in height and has arms that are 2·5 m (8 ft) from fingertip to fingertip. It weighs about 200 kg (440 lb). The female is about half this size.

● **Gorillas** rarely climb trees because of their size. A male can weigh 280 kg (620 lb),

Lar gibbon swinging through the trees. This form of locomotion is called brachiation.

Chimpanzee mother and young travel together years after the young is too big to ride. The relationship remains close until adulthood, at about twelve.

equal to four fully grown men. A single male, called a silverback, leads a family group of about a dozen members. Gorillas are generally gentle, quiet creatures. But if you meet one it is best to look down to show you are friendly; if you stare at a gorilla he or she will see it as a threat.

● A baby **chimpanzee** will travel some of the time on its mother's back until it is about five years old.

REPTILES

● Reptiles are cold-blooded animals that are usually slow-moving when cold, relying on the heat from the sun to warm them up and make them active. They have dry, scaly skin which does not allow water to evaporate from their bodies, so they can live in hot dry climates like deserts.
Most reptiles lay eggs with soft, leathery shells but some snakes and lizards give birth to live young.
There are about 6,500 different types of reptile. More than half of them are lizards.

TUATARA

There is no other reptile like the tuatara. While other reptiles are basking in the sun to get warm, the tuatara sleeps and comes out only at night, to feed on spiders and beetles. It is active at temperatures as low as 54°F, much cooler than any

The tuatara is found only in New Zealand. The female lays up to 15 eggs, which take 13-15 months to hatch.

other reptile. As it is so cool the tuatara moves and functions very slowly.

DID YOU KNOW ?

The tuatara breathes only once every seven seconds when it is moving, and once an hour when it is resting.

It grows slowly, taking 20 years to reach its adult length of about 60 cm (2 ft), and can live for over 100 years.
Species closely related to the tuatara existed before the age of the dinosaurs.

LIZARDS

● The **Komodo dragon** is the largest lizard in the world, reaching a length of about 3 m (10 ft) and a weight of over 100 kg (220 lb). This makes it longer than the floor to ceiling height of your room!
Komodo dragons live on the island of Komodo, Indonesia where their diet consists mainly of pigs and small deer, though they have been known to kill people. The Komodo dragon lies in wait for its prey and grabs it by the leg. It often does not kill prey outright, but the bacteria in its mouth causes the bite to fester very quickly. The prey cannot escape and becomes very weak enabling the dragon to tear it apart. It has been timed eating a goat in 10 minutes and a 45 kg (100 lb) boar in 15 minutes. It can eat more than its own weight in a single meal.

● There are only two types of poisonous lizard. The **gila monster** of Mexico and the **Mexican beaded** lizard.

● **Slow-worms** are legless lizards. You can tell the difference between a slow-worm and a snake by looking at the eyes; a snake's eyes are covered by a fixed transparent scale, but a slow-worm's eyes have moveable eyelids. Slow-worms are not poisonous. As with many lizards if you grab a slow-worm firmly by the tail, you will be left holding just the wriggling tail. Having shed its tail, it goes off and grows a new one.

● **Chameleons** can change the pattern and color of their skin to match the background trees. They swivel each eye around independently when hunting for insects. When one eye spots a victim, the chameleon focuses on it with both eyes, shoots out

A flap necked chameleon, found in Botswana, catches an insect, with its long sticky tongue.

its long sticky tongue catches the insect on the tip, and quickly coils the tongue up again into its mouth.

● **Geckos** are lizards that can run straight up a smooth wall or even across a ceiling. Their feet are covered with millions of tiny hooks that can cling to the slightest roughness of a surface.

SNAKES

● All snakes can swallow animals thicker than their own bodies because they have amazing jaws. The upper and lower half are linked by stretchy ligaments, and the lower jaw is made up of two bones also joined by elastic tissue. **Pythons** have the biggest mouths and can even get to grips with a whole goat, pig or antelope. A python will seize its prey in its mouth and then tightly coil itself round the animal's chest, not crushing it but preventing it from breathing. The animal suffocates. Once a python has had a massive meal it can do without food for several months.

● The **egg-eating snake** of Africa is about as thick as a man's finger but its favorite food are hens' eggs, which it gulps whole. Special sharp teeth at the back of its throat pierce the egg before it moves down the gullet.

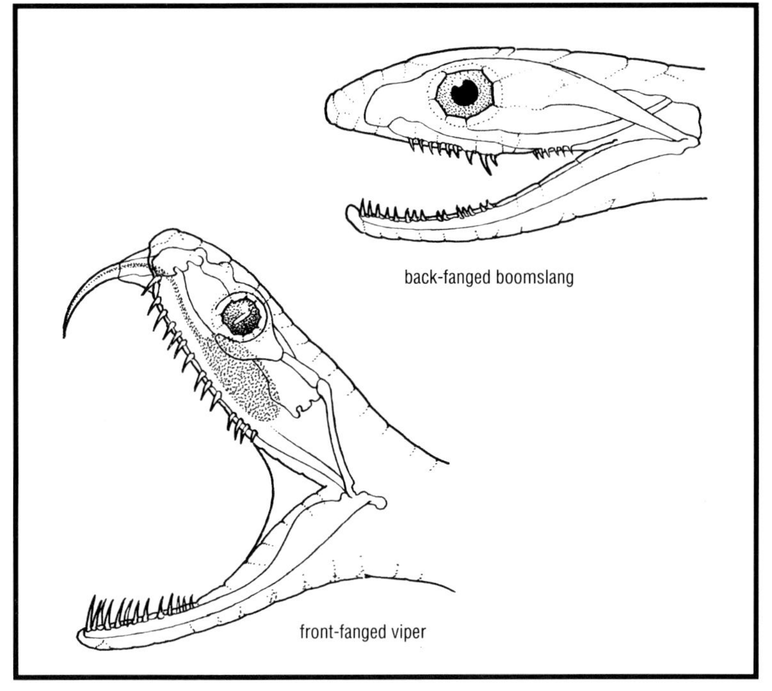

A third of all snake species are poisonous. Most have their fangs at the front of their mouths, but some like the boomslang, have them at the back.

● The **king cobra** is the longest of all poisonous snakes, reaching lengths of 5·4 m (18 ft). It injects so much venom in a single bite that it can kill an elephant in four hours. It is one of the few snakes to make a nest; the female coils on top of ther eggs until the young snakes wriggle out.

● The **spitting cobra** of Africa shoots venom out of its fangs aiming for its enemy's eyes. It is very accurate up to a distance of seven feet.

● The **African boomslang** snake has its fangs at the back of its mouth, not at the front like most other poisonous snakes. It has to get a firm hold of its victim in its jaws before the fangs can inject the deadly venom.

● Snakes do not have ear-drums and so cannot hear sounds, but they can sense vibrations such as footsteps. Cobras do not move to the rhythm of the music of the snake charmer's pipe but are probably mesmerised by its movements.

● **Rattlesnakes** can hunt in the dark because they are very sensitive to heat. They have two small pits just beneath their eyes which are crammed with heat sensitive nerve cells. These can detect a temperature rise of a few hundredths of a degree fahrenheit. We have heat sensing nerve cells on our skin, about 15 in each square inch: the snake has about 150,000 of them over the same area of pit organ. The rattlesnake can very accurately locate a small animal about a yard away. It then shoots its head forward at a speed of 10 feet a second, and its huge fangs inject its victim with a dose of deadly poison. Rattlesnakes are unusual snakes because they do not lay eggs. The female keeps the eggs inside her body until they hatch. The shell is only a thin membrane through which the

Diamondback rattlesnake in the Arizona desert.

baby snake receives some nourishment and oxygen from its mother's blood, in the walls around it. It also has a yolk sac for food, as in an ordinary egg. Once born the mother guards her babies by frightening away possible predators with her rattle.

● **Sea snakes** come to the surface of the sea to breathe but they can then stay under water for up to eight hours. They are able to do this because most of their body is lung; it even occupies space right into the tip of the tail. Sea snakes are about ten times more poisonous than the most poisonous land snake. They are found only in tropical waters.

● The **golden tree snake** of Malaya can leap from a tree onto its prey. Once launched it can glide for more than 25 m (80 ft). While in the air it keeps its body rigid and makes it into a hollow shape which traps air, rather like a parachute.

CROCODILES

● The saltwater, or estuarine, crocodile is the largest reptile. Found in southern Asia and northern Australia, it can grow up to 7.6 m (25 ft) and weighs almost 2 tonnes (2 tons). They can swim hundreds of miles

A bandy-bandy swimming through a coral reef in the New Hebrides. This is the only sea snake that lays eggs rather than giving birth to live young.

from land. They eat mainly fish and crustaceans, but will attack people.

● Crocodiles keep their eggs warm by laying them under a mound of earth or rotting vegetation. The temperature of the eggs is very important; if the temperature is below 86°F females hatch out, and if the temperature rises above 93°F males are produced. At temperatures between these two a mixture of males and females are produced. It is not known whether the crocodile is aware of this important information!

Alligators have shorter, blunter snouts than crocodiles. When a crocodile's mouth is closed, a large lower tooth still shows, fitting into a notch in the upper jaw. No lower teeth are visible when an alligator closes its mouth.

● The **Nile crocodile** takes very good care of its young. The eggs are buried under sand and as the young hatch out they make a loud piping sound that can be heard several yards away. The mother crocodile removes the sand covering her babies and puts them gently in her mouth. Carrying a few at the time she goes down to the water and swims off to a safe nursery area where she can keep a protective eye on them until they grow bigger.

TORTOISES AND TURTLES

● The **leatherback turtle** is the largest of the turtles. On average an adult grows to about 2 m (6 ft) long and weighs about 450 kg (1000 lb).

DID YOU KNOW?

The Aldabra giant tortoise is the largest tortoise at about 1.8 m (6 ft) long and weighing 225 kg (500 lb).

● The **alligator snapping turtle** of North America lies on the river bed with its mouth wide open, sticking out its brightly colored tongue. It tricks passing fish, who think this is a worm and swim into the open jaws of the turtle. This turtle is known to have a very good sense of smell. This was demonstrated when an American Indian helped police find dead bodies that had been dumped in lakes. He brought his snapping turtle on a long lead and put him in the lake. The turtle followed the smell of the rotting corpse and swam to it, looking for food!

The Galapagos giant tortoise grows up to 1·2 m (4ft) in length, and weighs up to 225 kg (500 lbs). It is an endangered species.

AMPHIBIANS

● Amphibians are cold-blooded creatures like reptiles and fish. Most amphibians spend part of their lives in water and part on land, and usually breed in water. Most young amphibians are usually water living and have gills for breathing. They undergo a process of complete change, called metamorphosis, which turns them into adults with lungs and, usually, legs, living on land. Even on land their skin is kept moist so that they can absorb oxygen through it, as well as breathing air through their lungs.
There are about 4,000 species of amphibian, divided into three groups: caecilians; salamanders and newts; frogs and toads.

CAECILIANS

● These legless amphibians live mostly in the moist ground of tropical forests. They look a bit like earthworms until they open their mouths to reveal powerful jaws, that can grab and devour small lizards. The largest can reach lengths of 1·5 m (60 in).

A caecilian in the rainforests of Costa Rica. There are 150 species of caecilian in all.

An adult axolotl. The axolotl, found at high altitudes in Mexico, is now endangered.

SALAMANDERS

● The largest amphibian is the **Japanese giant salamander** which can weigh up to 40 kg (88 lb) and grow up to 1·6 m (64 in) long: about the size of an average 12 year old boy.

● The **red salamander** has no lungs and only breathes through its skin, which must be kept moist. It lives on land and must keep out of the sun, where it could dry out and die within minutes.

● The **axolotl** is a salamander that never grows up. It was given its name by the Aztecs and it means 'water monster'. The axolotl stays a larva with external gills throughout its life. It can breed even though it is still a larva. Scientists have given axolotls growth hormone, and they have lost their gills, grown lungs and become land-living salamanders.

FROGS AND TOADS

● All frogs and toads blink at every gulp. They move their eyeballs down into the skull when they blink and this movement creates a bulge in the roof of the mouth, and helps squeeze food into the throat.

The marine toad is one of the largest toads. It can grow to more than 23cm (9 in) in length.

● **Goliath frogs** are by far the largest frogs in the world. A female has been caught weighing over 3·3 kg (7 lb). Her body measured 31 cm (12 in) long and her back legs were 43 cm (17 in) long, giving her an overall length of 74 cm (29 in). The goliath frog is so heavy that it cannot jump very well and it relies on its excellent rock-like camouflage to protect it from predators.

● Most frogs are able to take spectacular leaps. The record is held by a small North American frog called *Acris gryllus*. It is only 5 cm (2 in) long but can leap 36 times its own length, a distance of 1·8 m (6 ft).

● **Asian tree frogs** have special skin webs between their toes

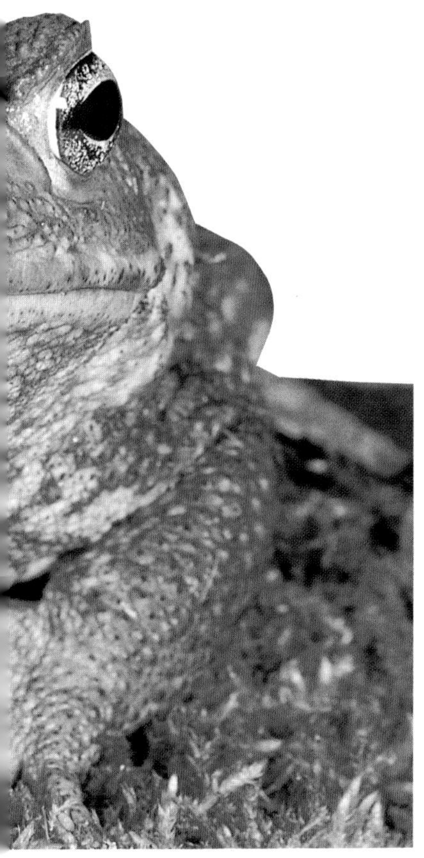

which act as parachutes as they leap and glide between trees up to 12 m (40 ft) apart.

● **Poison-arrow frogs**, found in the rainforests of South and Central America, have in their skin one of the most poisonous substances produced by any animal. It is used by Indian tribes for tipping their arrows when they go hunting.

DID YOU KNOW?

A single poison-arrow frog has sufficient poison to kill about 2,000 people.

● For a really protective father none can beat the **Darwin frog**. The female lays fertilized eggs on the ground and the male sits guarding them. As soon as the eggs show signs of life by moving slightly, he takes about a dozen of them into his mouth. He appears to swallow them but in fact keeps them in his throat in a sac that he normally uses to make loud croaking sounds. They stay there until one day he opens his mouth wide and out hop fully-formed froglets.

● Female frogs and toads can lay as many as a quarter of a million eggs in a lifetime. Only two need to grow up into adults for the population to remain constant. The vast majority of eggs are eaten by fish. Some frogs and toads lay fewer eggs and have unusual ways of protecting their developing offspring.

● Unlike any other toad the **midwife toad** mates on land and not in water. During mating the male fertilizes the eggs produced by the female, and then wraps the strings of eggs round his back legs. He carries the eggs and dips them in water occasionally to keep them moist, until the eggs hatch into baby toads.
The skin of the midwife toad is very poisonous and the liquid mucus on its surface can kill a mouse or adder on contact, although it is not strong enough to hurt humans.

● A small west African toad, only 2 cm (1 in) long, called *Nectophrynoides vivipara* gives birth to live young instead of laying eggs. The eggs are fertilized inside the female and develop there. They feed on small flakes shed from the inside of the female. After nine months the tiny toads are born.

● **Asian tree frogs** keep their eggs far away from hungry fish. The frogs breed on the branches of trees, choosing ones that overhang water. The

The male midwife toad carries up to 60 fertilized eggs in strings around his legs. He carries the eggs for between 18-49 days, making sure they do not dry out. He then leaves them in shallow water, where they hatch.

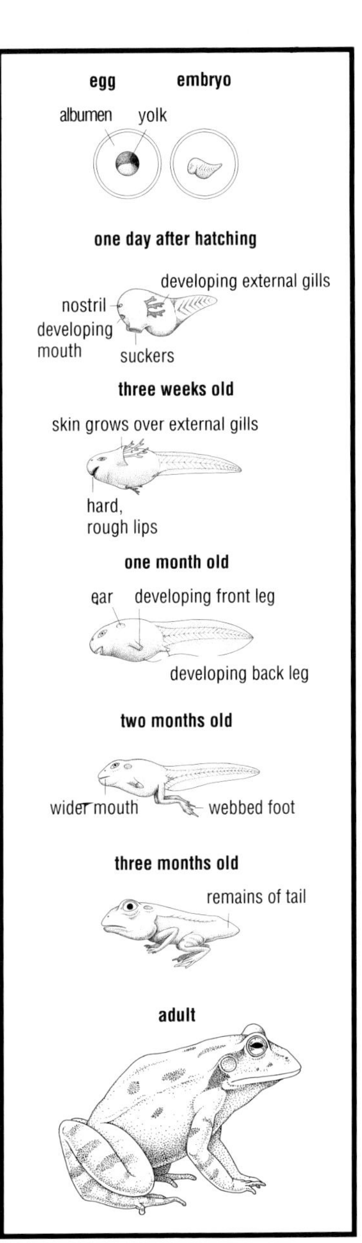

A frog hatches out as a tadpole, with gills for breathing and a tail for swimming. At first it uses yolk inside its body for food. Later it eats water plants, and finally it changes to a diet of small water animals.

male clings to the female's back to mate. As the female lays her eggs she produces a liquid which the male whips into a froth. The outside of this frothy mass hardens and keeps the developing tadpoles moist. When the froglets are fully formed they drop into the water beneath.

● **Marsupial frogs** have a pouch on their backs in which the fertilized eggs can grow. One species carries about 200 eggs at a time, releasing them into water when they have developed into tadpoles. Another species carries only about twenty young but releases them when they are fully formed froglets.

BIRDS

RECORDS

● The largest living bird is the **ostrich** which can grow to a height of over 2·5 m (8 ft) and weigh as much as 155 kg (340 lb). Although unable to fly, it can run at speeds of about 45 km/h (27 mph) and can even reach more than 70 km/h (43 mph) for short sprints. Its powerful legs can kill a man with a single kick. This giant bird is only supported by two toes on each foot.
The ostrich lays the largest egg. It is about 15 cm (6 in) long and weighs about 1·8 kg (almost 4 lb). Although the shell is only 1·5 mm (0·06 in) thick, it is very strong and will not crack under the weight of a 120 kg (266 lbs.) man.

● The smallest bird is the **bee hummingbird** of Cuba. It measures about 60 mm (2 ¼ in) long and weighs less than 2 g (0·07 oz). This makes it lighter than some hawkmoths.

● The smallest egg is laid by the **Vervain hummingbird** of Jamaica. It is about 10 mm (0·4 in) long and weighs less than half a gram (two hundredths of an ounce). This egg is laid in the smallest bird's nest in the world, so small that it would fit into half a walnut shell.

● A bird's feathers usually weigh more than all its bones. The bones are very light because some of them contain an open honeycomb structure filled with air. This means that

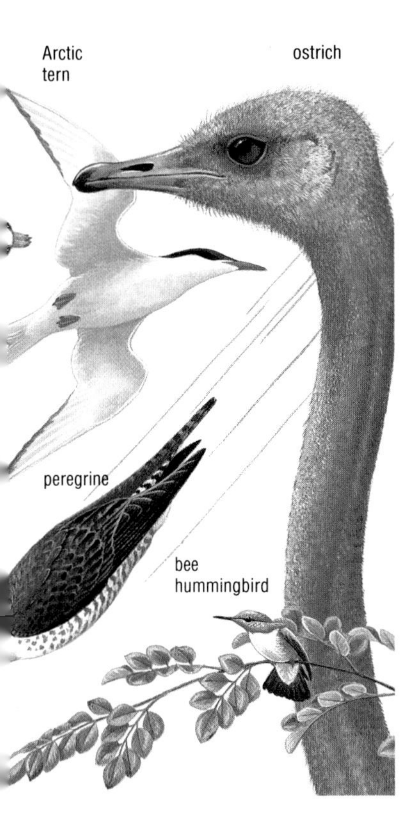

Some bird record-breakers: the ostrich is the largest bird and the bee hummingbird the smallest; the albatross has the largest wingspan; the peregrine is the fastest. The Greenland wheatear makes the longest non-stop flight (2,500 miles) and the Arctic tern the longest migration (24,800 miles).

a very large bird, such as the golden eagle with a wingspan greater than the height of a man, weighs only 4 kg (8·8 lb), about the weight of a good size newborn human baby.

● The heaviest flying bird is the **kori bustard** of Africa. It can weigh as much as 21 kg (46 lb) and although it can fly it prefers to run away from danger. Its long legs enable it to move quite fast.

● The largest wingspan belongs to the **wandering albatross**, reaching more than 3 m (10 ft). Its wing shape allows it to glide for hours over the ocean without a single flap of its wings.

● Tiny **hummingbirds** beat their wings at an amazing 50 times a second, and one type even flaps 90 times a second. This enables them to hover perfectly still while sipping nectar from a flower. They can also move backwards, upwards and downwards like tiny helicopters.

● The fastest flying birds are seen when the males are swooping and diving in courtship and territorial displays. Then the **white-throated spinetail swift** reaches a top speed of 170 km/h (100 mph) and the **peregrine falcon** has been recorded at a staggering

175 km/h (110 mph). Compare this with human sprinters who can run 100 m in 10 seconds, a speed of 36 km/h (22.5 mph). The fastest fliers in level flight, without swoops, are some kinds of ducks and geese. The **spurwinged goose** probably holds the record with speeds greater than 100 km/h (65 mph).

● There are about 100 billion birds in the world. A single type of African bird accounts for one tenth of all birds. It is estimated that there are about

Vultures feeding on a zebra carcass, Namibia. The vulture's bald head and neck mean it does not have to spend a long time cleaning its feathers after a bloody meal. Its strong, hooked beak is ideal for tearing flesh.

10 billion of the **red-billed quelea**. They darken the skies as they fly in flocks of hundreds of thousands.

FEEDING

● **Vultures** only feed on dead animals. They spend long hours looking for a meal hovering at considerable heights and scanning the landscape below for a carcass. The **griffon vulture** hovers at heights of up to 3,500 m (10,500 ft), and its amazing eyesight enables it to spot a dead animal several miles away. It is supported by its large wings, with a span of 2·8 m (9ft), as it glides on warm, rising air with no need to give a single flap of its wings for hours.

● The **bearded vulture** eats leftovers after other vultures have devoured the flesh of a

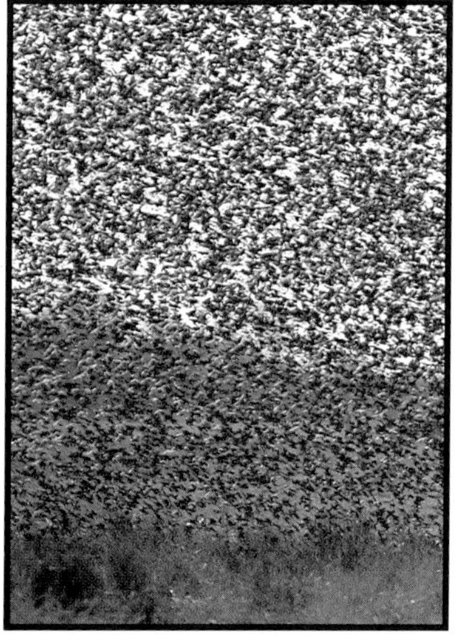

A flock of red-billed quelea takes to the air.

dead animal. Sometimes this animal only eats bones for weeks. It carries large bones to a great height and then drops them on rocks, smashing them into pieces small enough to eat.

● The **secretary bird** has hard scales to protect its long legs from the bites of the poisonous snakes it eats. This unusual looking bird stands about 1·2 m (4 ft) high and has a wingspan of over 1m (3.28 ft).

Pair of secretary birds, Kenya. When not breeding secretary birds spend most of their time on the ground. They kill their prey with their beaks, or by stamping on them.

● The **kiwi**, a flightless New Zealand bird, finds the insect larvae it feeds on by smell. It is the only bird with nostrils on the tip of its beak. All other birds have their nostrils at the base of the beak, where it joins the head.

● The **shrike** has a most unusual way of storing food. It catches voles, fieldmice and insects and then sticks them on the sharp spines of blackthorn. It is remarkable that this bird, which is only about the size of a starling, can successfully hunt mice.

● The **oxpecker** feeds on the little flies and ticks which live on the skin of large animals in Africa. This little bird can be seen riding on buffalo and antelope pecking at all the flies and parasites that live on the animal. It even feeds on those round the animal's cuts and scrapes. Scientists now think that a bird adopts a particular

Yellow-billed oxpeckers feeding on an African buffalo. Not only do oxpeckers stop their hosts from being troubled by ticks, they may also warn them of danger, as they give warning calls when alarmed.

A female Indian hornbill. Indian hornbills are the largest of the hornbill species. The male can grow up to 1·5 m (5 ft).

animal and returns to that one day after day.

● How can a bird get its favorite beetle larvae out of the little tunnels in dead wood where it lives? One type of **Darwin finch** living on the Galapagos islands has very cleverly solved this problem. It holds a cactus spine in its beak and uses it rather in the way that people use long pins to extract shellfish from their shells.

NEST BUILDING

● **Hornbills** make their nests in holes in tree trunks. Once the female is happily settled on the eggs the male builds up a wall in the front of the hole so she cannot get out. He leaves a small slit through which he provides food for her, and for the young once they are hatched. The female does not emerge till the young birds have feathers, and the male pecks the wall away. Many hornbills seem to form a strong attachment to each other and stay together for life.

● The **rufous ovenbird** of South America is sometimes called the potter bird because of the way it builds nests out of mud. Ovenbirds build large round nests on the top of posts or on tree branches. Inside there are two separate rooms and the whole structure is so solid that they can be used for years.

● **Lovebirds** carry nest-building materials in their feathers. These small African parrots do not just carry a single twig to the new nest but insert leaves and twigs among their feathers so they do not need to make so many journeys before the nest is completed.

● **Gannets** like to live in a crowd, and build their nests in large colonies of about 30,000 nesting pairs. Colonies of gannets off the coast of Peru even reach hundreds of thousands of birds. These enormous crowds of gannets produce a deafening noise and an unbearable smell!

A colony of cape gannets, Bird Island off West Cape, South Africa.

Adélie penguin chicks gather together for warmth.

DID YOU KNOW ?

The burrowing owl makes its nest by digging deep underground, to depths of about 1 ½ m (5 ft). It lives in desert regions in North and South America.

● **Adélie penguins** build a strange 'nest' in which they lay their eggs. On the ice they carefully arrange about 50 stones or pebbles in a circle that is just large enough for one of the parents to stand in. The eggs are laid and the parents share the job of sitting on them and keeping them warm against the bitter cold of the Antarctic. For the first two weeks the female sits on them, and then the male takes over until they hatch in about another three weeks. He does not eat during this time and loses about half his body weight.

● The young penguins huddle together in nurseries to keep warm, and their parents bring them food. No one is sure how the parents recognize their own young among so many others.

STRANGE BIRDS

● **Oilbirds** live in the pitch black of South American caves. They are very unusual among birds because they can find their way using echo-location, as bats do. They emit sounds which bounce off the surrounding surfaces, and the time it takes for these echoes to reach the bird tells them where they are. Unlike bats the sounds they produce can be heard by humans. At night they fly off and find fruit to eat using their well developed sense of smell. They are called oilbirds because the young are very fatty and are sometimes killed to make cooking oil.

● The **tinamou** of the South American rainforest just has not got the hang of how to fly. It will reluctantly take to the air if frightened and is then quite likely to crash into a tree and kill itself! Even if it has a clear path the tinamou will fly a short distance and then drop to the ground. Its heart is not large enough for it to do anything requiring too much energy.

Male highland tinamou hatching eggs. The eggs will have been laid by several different females.

The male mates with several females during the mating season. The females then lay their eggs in one nest and the male sits on all the eggs until they hatch.

● The **hoatzin** of South America is unlike any other bird. The young birds have little claws on the edge of each wing, which they use to climb through the trees. These claws disappear as the bird grows up. But the adult is not very good at flying and still uses its wings to help it climb. The hoatzin cannot fly at all after eating because its full stomach makes it topple forwards. It has to sit down until some of the food is digested.

The hoatzin builds its nest in trees overhanging water. The young birds can swim and if there is any danger, they can drop into the water to escape.

● The **turaco's** beautiful red color dissolves in the rain. No other animal in the world has a pigment which can be washed away. The green color of the turaco is also unique, as it has not been found in any other creature.

● **Puffins** can fly but cannot take off from level ground. They leap off the sides of cliffs and then flap their little wings to fly through the air.

LONG DISTANCE TRAVELERS

● The swift spends its whole life flying. It lands only to lay eggs and look after its young. The young birds leave the nest when they are about 7 weeks old and stay in the air for two or three years before returning to the nest site to lay eggs. They feed on tiny flying insects, and drink by skimming over water. They even mate in the air. At night they soar to a height of 2,000 m (6,500 ft) and sleep while flying, slowing the beating of their wings to 7 per second instead of their usual 10 per second. The swift covers about 900 km (560 miles) a day.

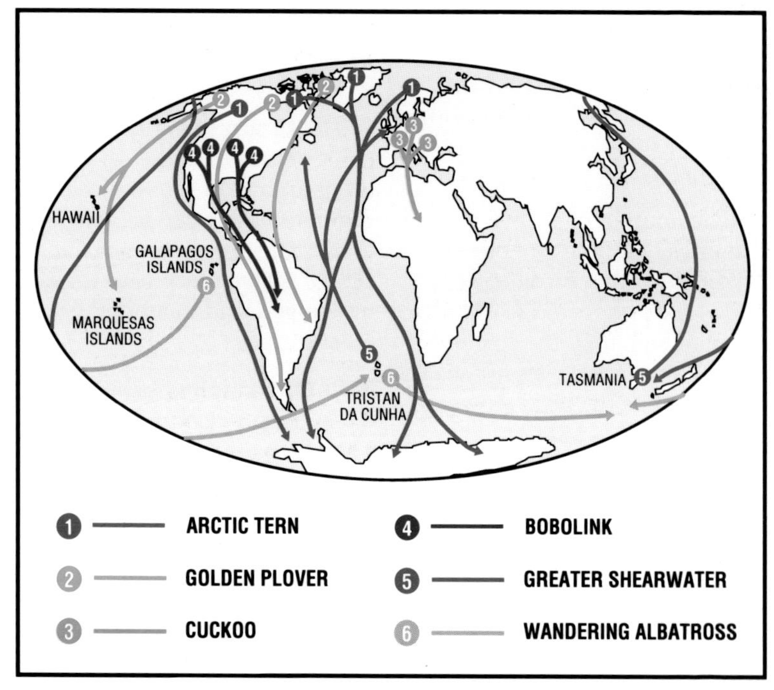

Map showing the migration flights of a selection of different birds.

● The **sooty tern** leaves its nest and does not return to land for three or four years.

● The **Arctic tern** is the greatest traveler of all. In July a chick hatches in its nest in northern Greenland. A few weeks later it takes off for the 18,000 km (11,250 miles) journey to the South Pole where it spends summer, returning the following spring to Greenland: a round trip of 36,000 km (22,500 miles).

● The American **golden plover**, a bird little smaller than a pigeon, flies non-stop from Alaska to Hawaii, a distance of 4,000 km (2,500 miles).

● Siberian **storks** and **ruffs** have to fly over the mountains of the Himalayas on their migration routes. They reach heights of over 5,500 m (18,000 ft).

FISH

● Fish are the largest group of vertebrates with over 20,000 different species. They are cold-blooded animals living in water, with gills for breathing. Their bodies are very muscular, and propel the fish with a wave-like movement through the water. Fish have fins which help their balance and in some species are also used for swimming.

● There are three types, or classes, of fish: jawless fish such as lampreys and hagfish; cartilaginous fish such as sharks and rays; and bony fish such as cod and trout.

● 95% of all fish species are bony fish with skeletons made of bone and smooth skin covered with bony scales. These fish have swim bladders which fill with air to keep them afloat.

● Cartilaginous fish do not have bones: their skeletons are made of a tough flexible material called cartilage or gristle. Their skin is rough and covered with sharp scales, and because these fish do not have swim bladders they sink to the bottom of the sea as soon as they stop swimming.

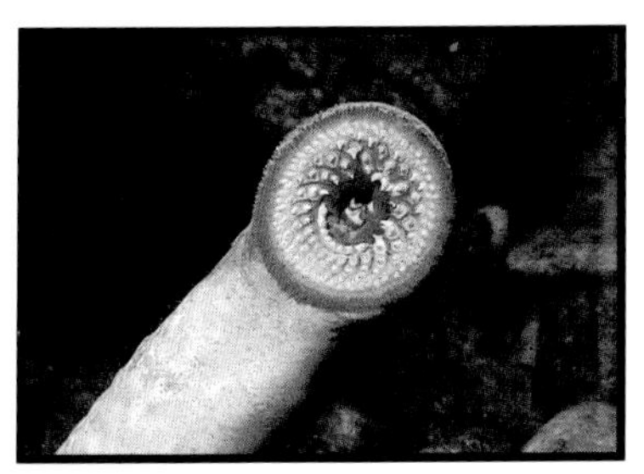

River lamprey (top); the sucker mouth of the sea lamprey (bottom).

JAWLESS FISH

● Lampreys and hagfish are jawless fish. **Hagfish** are slimy creatures living on dead and dying fish at the bottom of the sea. They bite a hole in the fish, slither in and eat it from the inside.
Lampreys have a rubbery sucker-like mouth. They fix themselves to live fish and push their tongue backwards and forwards like a piston until the sharp teeth on the tongue penetrate the skin of the fish. Lampreys then eat bits of flesh and suck the blood from the live victim.

SHARKS

● The **whale shark** is the largest fish and can grow up to 18 m (60 ft) long and weigh about 40 tonnes (40 tons). This giant shark only feeds on tiny creatures called plankton, which it filters from mouthfuls of water.

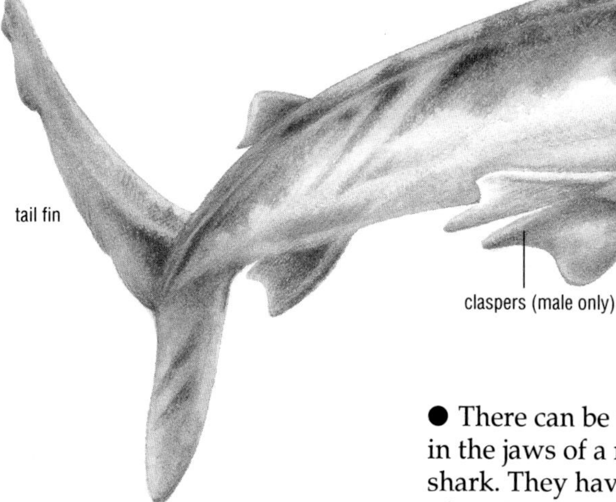

● Sharks have a very developed sense of smell and can home in on blood from over half a mile away. They also sense vibrations produced by struggling fish. Shark fishermen in the Pacific islands exploit this and rattle coconut shells underwater to attract them.

● There can be up to 3000 teeth in the jaws of a meat-eating shark. They have several rows of teeth in reserve, ready to replace any that fall out due to the stress of tearing the flesh of their prey.

● **Nurse sharks** shed their teeth and grow a new set every eight days. The teeth of the great white shark grow up to 5 cm (2 in) long and can tear off a person's leg with a single bite.

● Shark eggs are fertilized inside the body of the mother.

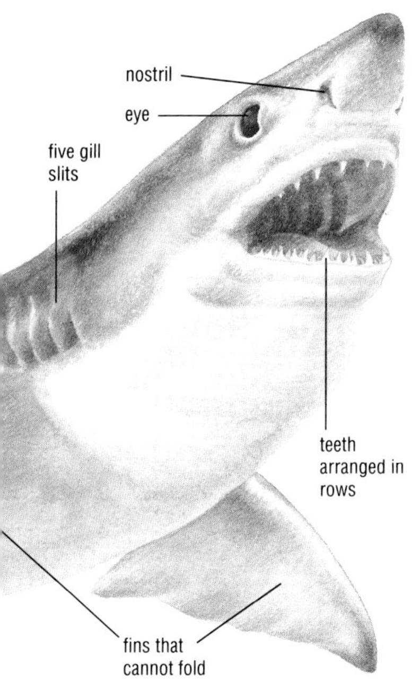

The great white shark can grow to about 6 m (20 ft) long. It is the most aggressive of all sharks and will attack humans.

In many types of shark the egg develops in a tough membrane called a mermaid's purse, and this package is deposited in the sea. In a few sharks the purse remains inside the mother and the young shark is born fully formed.

● The **mako shark** is one of the fastest fish, reaching speeds of 96 km/h (60 mph).

BONY FISH

● The **porcupine fish** is quite an ordinary looking fish until attacked. It then quickly swallows so much water that it becomes the size of a football. This startles the predator, and probably it makes it too large to eat.

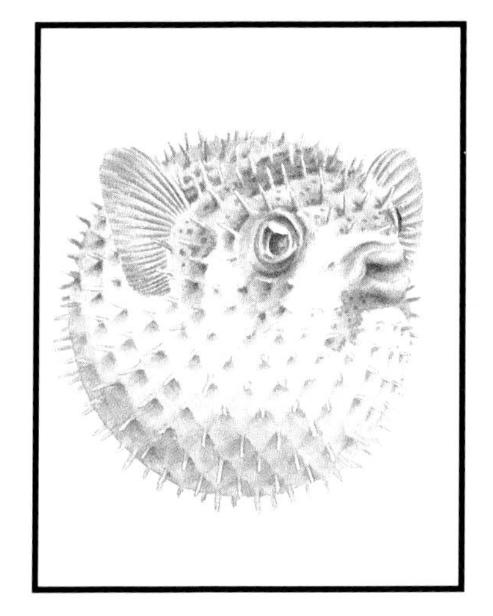

An alarmed porcupine fish.

DID YOU KNOW?

Sea bass (groupers) change sex as they grow older. They all start off as females but by adulthood some of them have grown male sex organs!

● In **seahorses** it is the male who gives birth. The male has a pouch on his body in which the female lays her eggs. There they are fertilized and develop until ready to be born. A few powerful contractions forces them out and they emerge, looking like smaller versions of their parents.

● Sole, flounder and other **flatfish** have an eye on each side of their head like ordinary fish, when they hatch. As they grow the body becomes flatter and the left eye moves across the top of the head, so that both eyes end up on the right side of the fish. The fish then swims on its side with both eyes looking up.

● **Stonefish** are the most poisonous fish in the world. They lie still and perfectly camouflaged among rocks and coral. If you were to tread on one of its spines the stonefish would inject you with a deadly poison.

● The deep-sea **anglerfish** lives in the complete darkness of the ocean depths. This could make finding food a big problem, but the anglerfish goes fishing! It has a spine sticking out of its head with a light dangling from it, in front of its gaping jaws. Fish attracted to the light 'bait' are eaten.
Finding a mate in the dark could also be a problem. But

Striped anglerfish dangles its worm-like lure as it 'fishes' for prey.

Clown anemonefish among the tentacles of their sea anemone host.

some types of anglerfish have solved this by having the male permanently attached to the female's body. The males, which are much smaller than the females, attach themselves onto the young female and actually become part of her. So the female does not need to search for a male because her eggs are fertilized by the attached male.

● **Archerfish** lie in wait near the water's surface, watching the plants on the river bank. When an insect settles on a plant the fish spits out a jet of water. This knocks the insect off the plant into the water, where the archerfish can snap it up.

● **Clownfish** lurk unharmed among the poisonous tentacles of sea anemones. Here they are safe from predators as any fish trying to catch the clownfish is hurt, and maybe killed and eaten, by the anemone.

● **Herring** swim in huge schools of several hundred thousand. The largest school was estimated to be made up of three million fish.

● Very few fish get a chance to hatch, as most eggs are eaten, and young fish also provide tasty meals for predators. Fish lay enormous numbers of eggs which ensures that one or two

The huge jaws of the gulper eel enable it to swallow creatures larger than itself.

survive. The cod holds the record with the female laying 6 million eggs at each breeding session. On a single day she can lay several hundred thousand eggs.

● **Flying fish** escape an attacker by jumping right out of the water and gliding through the air. They can leap as high as 6 m (20 ft) and cover distances of 300 m (1000 ft). The fish builds up speed under the water using its powerful tail and then breaks through the surface into the air. Spreading out its wide fins like

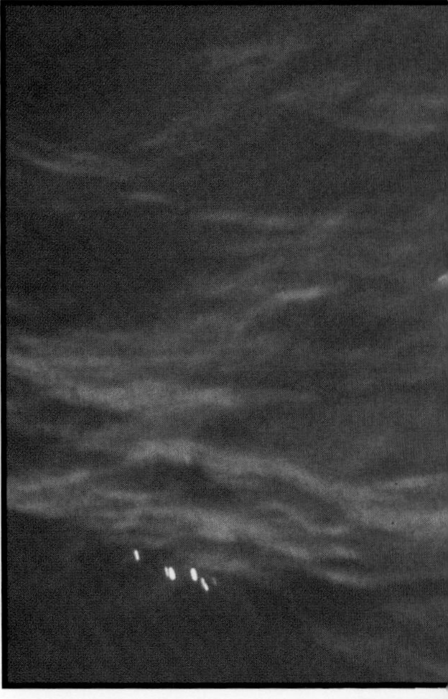

wings, it glides above the water, occasionally beating its tail against the surface for extra propulsion.

● **Gulper eels** live in complete darkness at depths of over 500 m (1,640 ft) in the Atlantic Ocean. Finding fish in the dark is not easy, so these eels have giant jaws which enable them to swallow any kind of fish, even some larger than themselves.

Four-winged flying fish gliding over the Red Sea. Flying fish feed mostly on other fish.

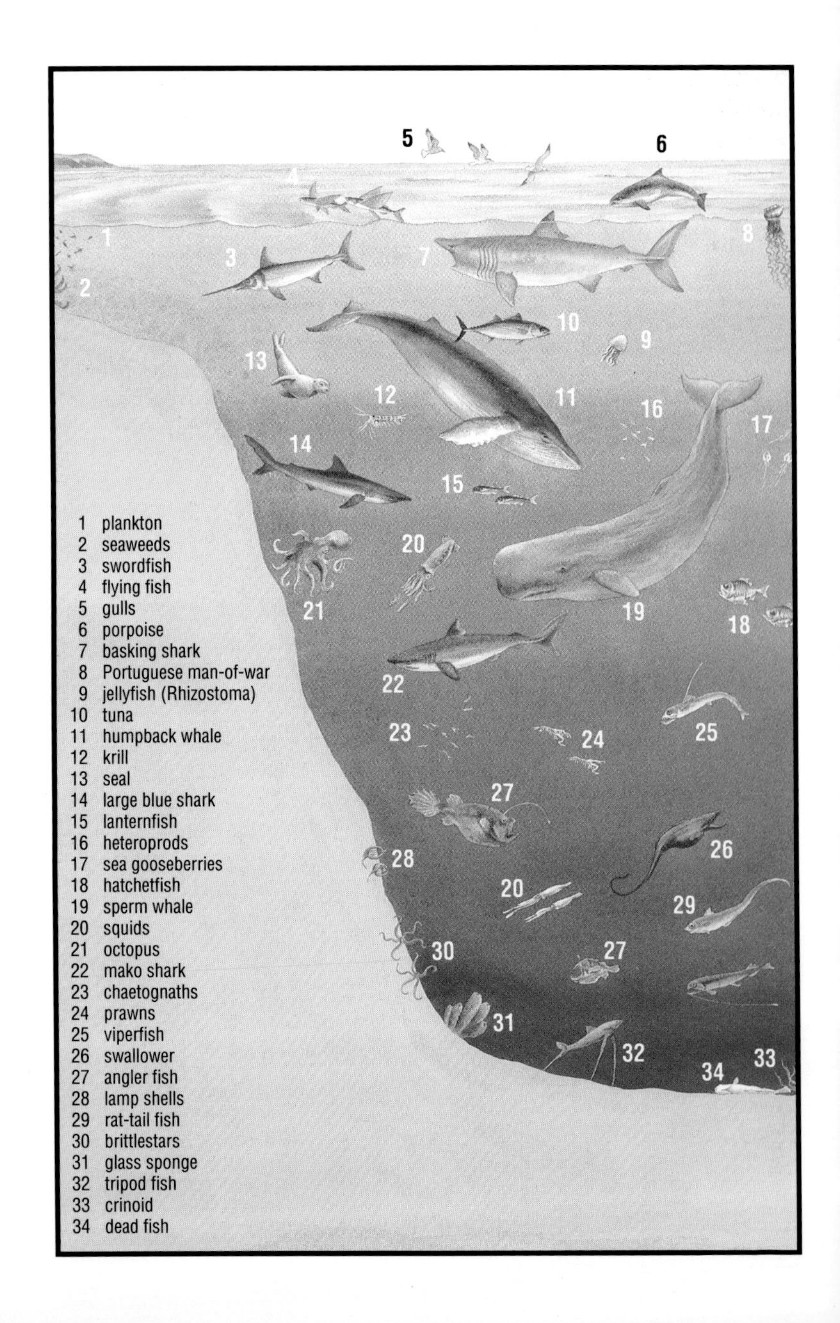
1 plankton
2 seaweeds
3 swordfish
4 flying fish
5 gulls
6 porpoise
7 basking shark
8 Portuguese man-of-war
9 jellyfish (Rhizostoma)
10 tuna
11 humpback whale
12 krill
13 seal
14 large blue shark
15 lanternfish
16 heteroprods
17 sea gooseberries
18 hatchetfish
19 sperm whale
20 squids
21 octopus
22 mako shark
23 chaetognaths
24 prawns
25 viperfish
26 swallower
27 angler fish
28 lamp shells
29 rat-tail fish
30 brittlestars
31 glass sponge
32 tripod fish
33 crinoid
34 dead fish

● The **electric eel** of South America can produce a shock of over 600 volts, which is enough to kill a person. This charge is released in a fraction of a second and kills or stuns any nearby fish, which are then devoured by the eel. Electric eels generate the electricity in their long tails. The eel is seven-eighths tail, and all its other organs are squashed up in the small area just behind its head. Living in dark, muddy water the eel cannot easily find its way around. Its eyesight and hearing are very poor, but it can use small electric discharges to 'see'. When swimming the eel produces about 30 electric impulses a second. This creates an electric field around the eel, and anything that gets in its path causes a change in the voltage. Such tiny changes in voltage can be detected, that this method is as sensitive as our eyesight.

● Adult **freshwater eels** leave European rivers every year for the Sargasso Sea, a stretch of ocean south-west of Bermuda. This is a journey of about 4,800 km (3000 miles) changing from freshwater to sea water. There each female eel lays millions of eggs and then dies. When the larvae hatch they return to Europe, a journey that can take them about three years. As they arrive at the coast of Europe the larvae change into young eels, called elvers, and swim up river.

A cross section of part of the ocean showing the depths at which fish, and other sea creatures, live.

DID YOU KNOW?

Moray eels have a remarkable sense of smell; much more sensitive than even that of a highly trained tracker dog.

● The **coelacanths** were thought to have died out 70 million years ago. Only a few fossil remains of these creatures had been found until 1938 when a live specimen of the fish was caught in the nets of fishermen off the coast of South Africa. Since then several have been caught. These fish can grow up to 1·5 m (5 ft) long and weigh over 50 kg (120 lb). They have unusual fins which are rather like little legs. The coelacanth gives birth to live young.

OTHER SEA CREATURES

● A natural bathroom **sponge** looks like a slimy piece of raw liver when alive. This extraordinary animal is basically a hollow sac with holes in the side and a larger opening at the top. Cells on the inner surface have little tail-like structures which constantly move backwards and forwards, forcing water out through the top. Water moves in through

Red vase sponges growing in amongst coral. Many tropical sponges are brightly colored, and some are as large as 2m (6ft) in height.

the holes in the side bringing oxygen to all the cells. The water also contains tiny food particles, which get stuck to special cells that wrap themselves round the particles and digest them. Other cells move freely through the sponge collecting digested food and moving it to other parts of the sponge.

● To make a bathroom sponge the cells are boiled away so only the skeleton remains.

● Some **corals** live by themselves, but many live in colonies made up of separate units called polyps. A living polyp looks like a little sea anemone, with tentacles to trap tiny animals for food. The polyps also have minute algae plants living inside them. The algae, like other plants, produce their own food by photosynthesis. During photosynthesis oxygen is released, so the algae help provide the polyps with oxygen.

● In stony corals the polyp secretes a limestone skeleton which links up with neighboring polyps. When the polyps die the hard skeletons remain, building up into a coral reef. Only the surface of the coral is alive, made up of millions of polyps, each only a few millimetres across.

● The Great Barrier Reef of Australia is a coral reef extending over a thousand miles. The formation of this amazing structure has taken hundreds of millions of years.

● Coral reefs grow only in waters where the temperature never falls below 21°C (70°F). Other small single corals are found in cooler waters, some even grow in the deep cold waters of Norwegian fjords.

● **Fan worms**, **peacock worms** and **feather worms** all live in long tubes which they build in the sand. Only the head end of the animal sticks out. It has long feathery gills which absorb oxygen from the water, and also trap tiny particles of food pushing them into its central mouth. They usually have eyespots on the gills and the slightest shadow passing over them will make the worm pull its tentacles into the tube at lightning speed.

● The **Portuguese man-of-war** is a relative of the jellyfish. It has tentacles over 12 m (40 ft) long, each with thousands of stinging capsules containing a poison as dangerous as that of a cobra. The Portuguese man-of-war looks like it is a single

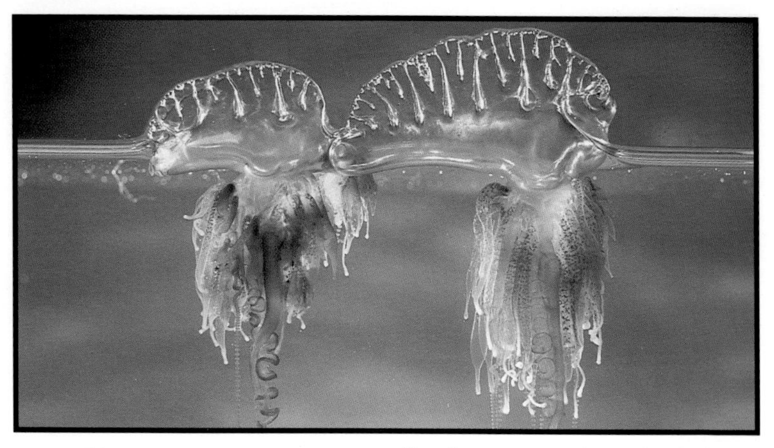

These Portuguese men-of-war, called blue bottles, are found in Australian waters.

animal but it is in fact a collection of different animals, called polyps, living together in a colony. The polyps are joined together in long chains to form tentacles, and each is unable to survive alone. The polyps perform different functions including catching food, digestion, and reproduction.

● The largest jellyfish is the **Arctic jellyfish**. It can grow up to 3·6 m (11 ft) across with tentacles 35 m (115 ft) long.

● 95% of a jellyfish is water, and if they get stuck on a beach they just melt away within a few hours, leaving a damp stain in the sand.

● The **box jellyfish**, also known as a sea-wasp, contains one of the most powerful poisons in nature and it can kill a person in a few minutes. It is found in the waters off the coast of Australia and South-East Asia. It is only about 25 cm (10 in) long.

● A **sea cucumber** protects itself from predators by ejecting its sticky inner organs over fish and crabs that start poking it. The sea cucumber then slowly moves away leaving its guts behind. It will grow more in the next few weeks.

The sea cucumber has an opening at each end of its body. One end, surrounded by tentacles for catching food, is its mouth. The other end is used not only for excreting but also for breathing. It is through this opening that a most

Red-striped sea cucumber, Papua New Guinea. Sea cucumbers are echinoderms, 'spiny skinned' creatures, related to starfish.

unwelcome visitor enters the sea cucumber. A slender pearlfish can sneak inside this hole and shelter there. The pearlfish will sometimes start eating the sea cucumber from the inside.

● **Starfish** usually have five arms but different types can have almost any number between four and fifty. The arms are extremely powerful and can force open shellfish such as oysters, scallops and mussels. Starfish will even tackle clams, fixing their feet on either side of the two halves and very slowly prising them apart. It then pushes its stomach into the gap and digests the flesh inside the shell.

Some starfish can reproduce in a most peculiar way. They shed one or more arms and each arm grows into a new starfish. The

A common Florida starfish about to attempt eating a flame scallop.

parent grows new arms to replace those it has lost.

SHELLFISH

● The **scallop** is a most unusual shellfish. It has eyes all round the edge of its shell. These are not merely basic light detectors: they have a lens, cornea, iris and retina, similar to our own eyes. Scallops keep watch for approaching predators such as starfish. Once a starfish has been spotted the scallop, unlike most shellfish, can make a rapid escape by opening and closing its two hinged shells and squirting out water: a type of jet propulsion.

DID YOU KNOW ?

The giant clam is the largest shellfish. It can grow to over a metre (39 in) across and weigh more than 250 kg (over a quarter of a ton).

● All **oysters** start life as males. They slowly change into females, lay eggs and then change back into males. They make this sex change about once a year in cold waters, and more frequently if living in warm waters.

● Thousands of dollars are paid by collectors for rare

specimens of **cone shells**. These beautiful sea snails can be deadly when alive. They have a long ribbon-like tongue covered with dart-shaped teeth. A cone-shell sticks out this tongue, called a radula, in the direction of its prey, which may be a worm or even a small fish. It then shoots out a tiny harpoon which spears the animal and keeps it tethered while poison is injected. This poison is so powerful it instantly kills a fish and can even kill a human. The lifeless prey is drawn back towards the cone shell where it is slowly digested.

Elephant trunk cone shell on the Australian Great Barrier Reef.

SQUID AND OCTOPUS

● Most squid are only an inch long but some real giants live in the depths of the north Atlantic ocean. The largest specimens which have been seen have bodies over 3 m (10 ft) long, with tentacles over 15 m (50 ft). But some believe that lurking deep in the ocean are even larger squid. The evidence for this comes from sperm whales who plunge the depths in search of squid to eat.

Many squid have special organs that enable them to produce their own light.

In the battle for its life the squid leaves sucker marks on the whale. A 15 m (50 ft) squid has 10 cm (4 in) suckers, but what size squid leaves the 45 cm (18 in) sucker marks that have been seen on some whales? Could there really be squid with a body twice the height of a house and tentacles 60 m (200 ft) long, or were the marks made by something else?

● **Giant squid** have the largest eyes of any animal. Each of their two eyes measure 40 cm (16 in) across, bigger than a large frying-pan. Their eyes seem to be better than our own as they have twice as many light sensitive cells on each square inch of the retina, so they probably can see much more detail than we can. Squid have large brains with which to process all this visual information, and they have very fast reactions.

● **Luminous squid** have an elaborate system of flashing lights. The organs which produce the light have lenses, mirrors, and diaphragms to control the amount of light leaving, and shutters to block it off completely. One type of squid has searchlights which it can move backwards and forwards, and up and down, tracking prey. It will also use the light to blind its prey. The diademed squid has the most spectacular display of colored lights. Their bodies look like they are trimmed with sparkling precious stones in ruby-red, bright blue and vivid white.

● **Octopuses** have large brains and a well-developed nervous system. They can easily be trained to distinguish between different shapes such as a cross

The highly venomous Australian Southern blue-ringed octopus.

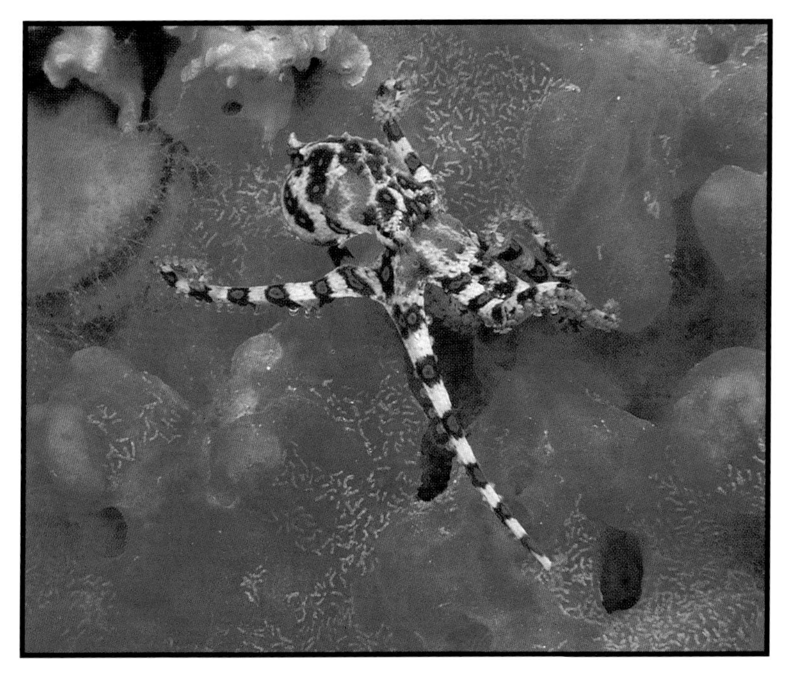

The giant cuttlefish can grow up to 2m (6ft) in length. Cuttlefish have eight legs, and two feelers round the mouth.

and a square. They also have good color vision and can change their body color to match their surroundings.

● Octopuses are a favorite food of the moray eel. The octopus releases a jet of murky liquid to escape from the eel. The liquid produces a strong smell which confuses the moray eel's sense of smell.

DID YOU KNOW?

Octopus can grow to 10 m (33 ft) across.

● The blue-ringed octopus is only 3 cm (1 ¼ in) long but can bite and kill anyone who picks it up. Its poison is more deadly than that of any land animal.

● **Cuttlefish** can change color rapidly to match whatever

background they are swimming over. In their skin there are small bags of red, orange, yellow, brown and black pigments. Muscles contract or expand these bags producing a wide range of colors.

CRUSTACEANS

- Crustaceans have a hard crust-like shell protecting their bodies. They include crabs, lobsters and shrimps, and range in size from the tiny water flea, less than 0·25 mm (0·01 in) long, to the **Japanese spider crab** which can measure 3·7 m (12 ft) across the span of its claws.

- When a crab or lobster grows, its hard shell cannot expand so it must be discarded and replaced by a new one. Before shedding its shell the animal absorbs into its blood calcium carbonate from the shell, and then produces a new soft wrinkly skin beneath the shell. The shell splits open and out emerges a soft animal. Its soft skin absorbs water, swells and slowly hardens.

- **Hermit crabs** never produce a hard shell of their own. They adopt an empty sea shell as protective armor. As the crab grows it moves in to a new, larger shell. Sometimes sea anemones take up residence on the shell and feed on the crab's leftovers.

Hermit crab with an anemone living on its shell, Devon, England

● **Horseshoe crabs** clamber onto dry land for three nights in spring when there is a full moon and high tides. Usually they live in the depths of the sea off the coasts of south east Asia and North America. In spring they move to the coast for night-time mating. The females move onto the sand, dragging one or more smaller males behind. The females lay eggs in the sand and the males release sperm. It is an amazing sight: a seething mass of thousands of crabs stretching for miles along the water's edge. Horseshoe crabs are actually more closely related to spiders than crabs.

Horseshoe crabs laying eggs on a beach in New Jersey, USA.

● The largest lobster is a North American species called *Homarus americanus*. It can grow a metre (3 ft) long and weigh 18 kg (40 lb).

● Lobsters like eating all sorts of sea creatures, dead or alive, but especially dead. They are also cannibals and will devour other lobsters, smaller than themselves.

● **Krill** are tiny shrimp-like crustaceans eaten by some whales. They live in the cold waters around the North and South poles, feeding on all sorts of microscopic plants and animals called plankton. Krill swarm in enormous schools and whales plow through them with open jaws, taking great mouthfuls. The krill are trapped in the filtering system in the whale's mouth. The blue whale eats four tonnes (4 tons) of krill a day.

INSECTS & OTHER LAND ARTHROPODS

● There are about 900,000 species of insects and thousands of new species are discovered each year. All insects have a body that is divided into three segments: head, thorax, and abdomen. Attached to the thorax are three pairs of jointed legs and usually two pairs of wings. Insects are invertebrates which means that they do not have a backbone. Instead of an internal skeleton they have a hard outer skeleton. Several other groups of invertebrates also have hard external skeletons and jointed legs: creatures such as spiders, scorpions and centipedes. They and the insects all belong to a large group of animals called arthropods. Arthropods include:

insects (ants, termites, bees, butterflies, beetles)

chelicerata (spiders, scorpions, mites)

chilopoda (centipedes)

diplopoda (millipedes)

● It is estimated that there are one billion billion (1,000,000,000,000,000,000) insects in the world. This means that for every human

An insects body is made up of three different sections: the head, thorax and abdomen. Insects have three pairs of legs attached to the thorax and many adults also have two pairs of wings.

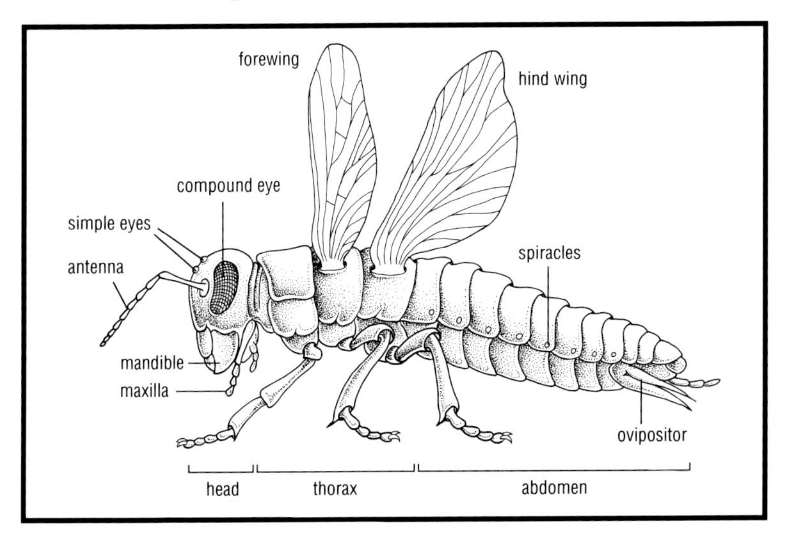

there are about one million insects. The total weight of insects is about twelve times the weight of all the people on this planet.
So far almost a million different species have been identified and named. There are probably another 5 million different types of insects waiting to be identified. At present 7,000 new species are discovered each year.

● The heaviest insect is the **goliath beetle** which can weigh up to 100 g (3 ½ oz) with an overall length of 11 cm (4·3 in). They are found in the equatorial region of Africa.

● The longest insect is the **giant stick insect** found in Indonesia. It can grow up to 33 cm (13 in) long.

● The smallest insects are tiny wasps, called **fairyflies**, which live as parasites in the eggs of other insects. They are less than a quarter of a millimetre (0.01 in) long and their eggs weigh about five thousandths of a milligram (so 140 million eggs would weigh only an ounce).

SOCIAL INSECTS

● **Ants** live in very well organized groups or colonies. Each member of the colony has her own particular job and by working together ants are able to use plants and animals in most unusual ways.

● **Weaver ants** of south east Asia build a nest by sewing leaves together. Groups of ants pull two leaf edges together and fix them using silk to bind the edges firmly together. The silk is produced by gently squeezing a young ant larva and moving it backwards and forwards until the leaf edges are stuck together.

● African robber or **slave-making ants** force ants of other species to do all the work for them. These slave ants collect food and feed it to their mistresses.

● The **honeypot ants** of Australia collect nectar and feed it to a special group of worker ants. These workers are fed so much that they swell up until they are as much as a half inch across, and are too big to move. The other ants hang them up in underground chambers and use them as living storage pots of food.

● **Army ants** are really terrifying animals. They do not have nests, but march through the countryside looking for animals to eat. They march in a

Australian weaver ants use larvae as silk shuttles, to join leaves together to make a nest.

column so long that it can take several hours for it to pass by. At the front and sides of the column are the soldier ants, totally blind and with huge, menacing jaws. Behind the soldiers come the workers, some carrying the young ant larvae. When the soldiers find an animal in their path all the ants swarm and devour it. They will feed on anything that does not escape in time: an aphid, a young bird, a python or even a horse tied to a post. After weeks of marching the ants make 'camp'. They form a large ball of hundreds of thousands of ants all clinging together. This 'living' nest has passages and chambers where the queen can lay her eggs, which sometimes number up to 300,000. When new larvae have hatched the army marches off again.

● **Leaf-cutting ants** from South America work day and night collecting leaves and stems from trees. They cut them into tiny pieces to carry back to their underground nest. Here they chew them and use them as a compost on which to grow a fungus. It is the fungus which the ants eat. These ants can rapidly demolish a tree: a fruit tree can be stripped of its leaves overnight.

Termite mound in Kakadu National Park, Australia.

● **Termites** build amazing mounds which can reach 6 m (20 ft) high and contain more than a million termites, all of them the offspring of one female (the queen), and one male (the king). The queen can live for up to fifteen years. Her enormous, swollen body, which is about 12 cm (5 in) long, produces 30,000 eggs a day.

The termite mound is an amazing piece of insect architecture. It is constructed from up to ten tons of mud and has special channels in it to provide ventilation, and keep fresh air circulating throughout the mound.

● **Bees** live in hives which contain one egg-laying female (the queen), a few hundred stingless males (drones) and about 50,000 females (workers) that are sterile and cannot lay eggs. The queen makes a single flight from the hive; the drones go with her and mate with her in flight and then die, or are later killed by the workers. For the next three or four years the queen will lay at least a 1,000 eggs a day throughout the summer months. All of these eggs were fertilized during her single mating flight.

A queen honey bee (center) being looked after by her workers, whilst she lays eggs.

BUTTERFLIES AND MOTHS

● Male butterflies and moths have an amazing sense of smell and can detect a female several miles away.

DID YOU KNOW?

The male emperor moth has the most powerful sense of smell in nature. He can smell a female about 11 km (7 miles) away.

● **Monarch butterflies** go on very long migratory flights each year. They live in North America and each autumn they fly south 2000 km (1,250 miles) to reach warmer places. They rest during the winter in trees in Mexico, and in spring fly all the way back home.

● **Morpho butterflies** have wings of beautiful shimmering blues and violets, but their wings do not have any blue pigment. Each wing is covered with scales as in other

Rajah Brooke's birdwing butterfly, Malaysia. The birdwings are the largest butterflies.

butterflies, but each scale has over 1,400 microscopic ridges. Each ridge splits the light that strikes it and reflects only blue. This is called irridescence and can also be seen in hummingbirds and some ducks.

● **White butterflies** have been found well inside the Arctic Circle, further north than any other butterfly. They are also found at heights of 5,000 m (16,000 ft) in the mountains of Africa and Asia.

● The largest butterfly in the world is the **Queen Alexandra's birdwing** found in parts of Papua New Guinea. The female has a wingspan of over 28 cm (11 in) and weighs about 25 g (0·9 oz).

● The smallest butterfly is the tiny ***Micropsyche ariana*** with a wingspan of 7 mm (¼ in), less than the size of the nail of your little finger.

BEETLES AND FLIES

● **Fireflies** are beetles which produce flashing green lights. They use the lights to signal to members of the opposite sex. Each species of firefly has its own code of flashing on and off. Each flash lasts an exact length of time and there are regular intervals between the flashes. This means that they do not attract beetles of different species. But some predatory beetles imitate the flashes of another species to attract them as prey.

● When attacked **ladybugs** squirt out some of their foul-smelling blood as protection.

● **Dragonflies** are the fastest insects in the world; they can reach speeds of over 50 km/h (31 mph). Small insects, such as mosquitoes, beat their wings much faster than dragonflies but only achieve speeds of about 2 km/h (1 mph). The rapid movement of their wings produces a buzzing noise.

Insect	Wingbeats per second	Maximum flight speed (mph)
Dragonfly	25-40	34
Cockchafer beetle	50	7
Hawkmoth	70	30
Bumblebee	130	7
Housefly	200	4
Mosquito	600	1
Midge	1000	½

● **Tsetse flies** are blood-sucking flies that carry the deadly disease of sleeping

Tsetse fly feeding on a human. There are about 20 kinds of tsetse fly and most of them will attack people.

sickness to both humans and animals throughout Central Africa.
The tsetse fly reproduces in a most unusual way. The female does not lay eggs but produces just one egg which she keeps and feeds inside her body. This young larva breathes through two tiny black tubes which stick out of the mother's body. When ready to pupate and turn into an adult the mother gives 'birth' to the young larva which she squeezes out of her body onto the ground. A female tsetse fly produces about twelve larvae in the six months that she lives.

● **Scarab beetles** provide their offspring with an unusual food. The male and female beetle roll the dung of animals such as buffalo, into balls, often much bigger than themselves. The male digs a hole and the two parents push the ball of dung into it. The female lays eggs on top and the developing larvae feed on the dung.

● When **burying beetles** smell the scent of a dead animal they fly off to the corpse. With animals about the size of a bird the beetles bury the body by removing the earth from underneath. They feed on the animal and the female lays her eggs on the flesh. The body is completely covered with earth, and when the eggs hatch they feed on the fleshy remains.

Scarab beetle rolling a ball of dung. The dung provides food for the scarab's larvae when they hatch.

● The **oil beetle** larva needs a lot of luck to grow up into an adult beetle. The eggs are laid in the ground and the newly hatched larva climbs up a plant stem and waits on the flower petals for the arrival of a bee. When the bee comes to collect nectar the tiny beetle larva attaches itself to the bee and is taken back to the bee's hive. There it climbs into a cell, eats the egg and pretends to be a bee larva, feeding on the nectar brought by the bees. These beetles are called oil or blister beetles because they produce an oily, poisonous liquid that can cause nasty blisters on the skin.

● The adult **ant-lion** looks like a dragonfly but it is the behavior of its larva that has given this creature its name. The larva lives in sandy ground where it carefully makes a steep-sided pit in the sand in the shape of a funnel.

South African ant-lion larva burrowing backwards into the sand. When it has dug a pit it will wait at the bottom for a small insect to fall in. Seizing the insect with its powerful jaws the ant-lion will suck it dry.

Any passing ant which stumbles into the funnel cannot get out and is grabbed by the powerful jaws of the ant-lion, and devoured.

SPIDERS, SCORPIONS AND MITES

● **Silk spiders** weave webs up to several yards across, and they wrap their eggs in a cocoon of silk which is made up of nearly a mile of silk thread.
The female silk spider is ten times larger than her male partner. She measures about 15 cm (6 in) from leg tip to leg tip, whilst he measures only 1·5 cm (about ½ in), and weighs only a

A Kenyan trapdoor spider lifts its trapdoor and emerges from its burrow.

hundredth of her weight. These spiders are found mainly in equatorial regions.

● **Bird-eating spiders** of South America are the largest spiders in the world, with a body size of about 9 cm (4 in) across, the width of a man's hand. Their legs measure 26 cm (10 in) across. These enormous spiders devour other spiders, lizards, small mammals and baby birds. They can live up to 17 years.

● **Trapdoor spiders** live in holes underground and, to protect themselves from intruders, they make a cover for the opening of the hole. This trapdoor is shaped by the spider so that it fits perfectly. It is even camouflaged with a layer of moss. The spider makes two tiny holes on the underneath of the door so that it can hold it in place and keep it firmly shut.

● The **scorpion's** venom is injected by a sting in its tail. A single sting can kill a small rodent in a few minutes.

● **Mites** are found in all places on our planet, and on all parts of different animals. One type lives in your eyelashes, another species lives in the nostrils of snakes, and another in cheese. They have even been found in the freezing north, and in scorching deserts.

CENTIPEDES AND MILLIPEDES

● **Centipedes** usually have between 15 and 23 pairs of legs, although their name means 'hundred legs'.

A female scorpion carrying her young on her back. All scorpions produce live young, rather than eggs. The young travel on their mothers back for several days.

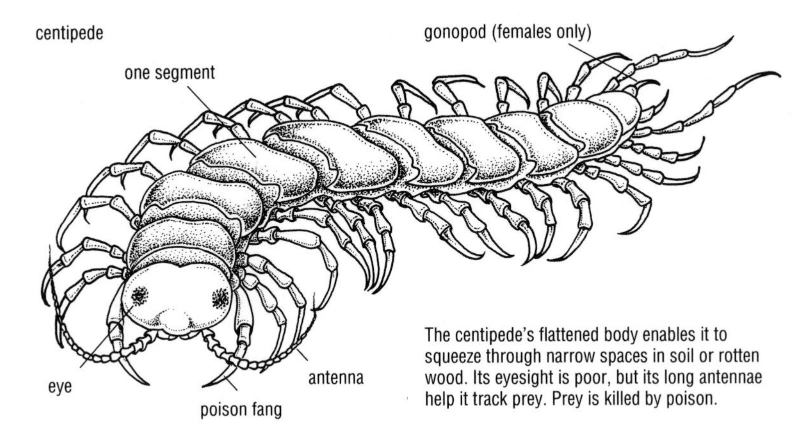

The centipede's flattened body enables it to squeeze through narrow spaces in soil or rotten wood. Its eyesight is poor, but its long antennae help it track prey. Prey is killed by poison.

In Central and South America centipedes can grow up to 30 cm (12 in) long and can give very painful bites which have been reported to kill people. Their main foods are locusts, cockroaches and lizards. Centipedes are long-lived and may live for about six years.

● **Millipedes** have a name that means 'one thousand legs' but they usually have less than a hundred. Unlike centipedes they are vegetarian. The longest millipede can grow to 30 cm (12 in) and is found in the Seychelles. The smallest millipedes are as little as 2 mm in length. Some species of millipede live up to seven years.

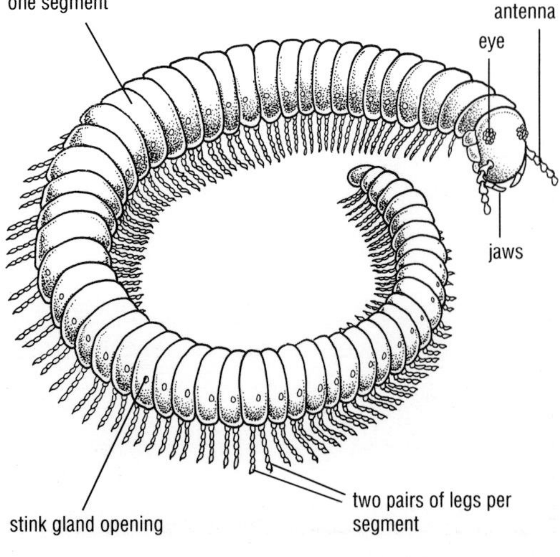

Millipedes have a row of stink glands along their sides, from which they produce unpleasant chemicals when disturbed. Their legs provide enough power to enable them to burrow in search of decaying plants to eat.

ENDANGERED SPECIES

● Every day another species of plant or animal becomes extinct. Many of them are eliminated by human actions. Their habitats are destroyed as natural resources are plundered, and many animals are hunted relentlessly because some people find them more attractive dead than alive. Elephants are hunted for their ivory tusks, used in ornaments and jewelery; many animals, including the big cats, are hunted for their fur to make coats.

● In 1970 the **black rhinos** of Africa numbered 65,000 but now there are less than 3,000. They have been hunted for their horn which is used to make dagger handles in the Middle East, and as a love potion.

● Today there are 625,000 **African elephants** but between 1979 and 1989 70,000 a year were slaughtered for their ivory tusks. At that rate they would have been extinct before the end of the century.

The giant panda is extremely rare. Its rarity is mainly due to changes in climate and vegetation since the end of the last ice age, rather than due to human activity.

● The table shows a few of the species close to extinction. But it is not only animals whose numbers are very low that are at risk. It does not take long to wipe out a well-established species. Two hundred years ago the **passenger pigeon** flew across the North American skies in huge flocks of millions. They were determinedly hunted and eaten by the settlers. In 1914 the last passenger pigeon died in a zoo.

DID YOU KNOW?

Today the International World Conservation Union has listed over 4,500 animal species in danger of becoming extinct. There are over 500 mammals and 1000 bird species on that list. This is about one eighth of all known bird and mammal species.

Animal	Location	Approximate number left
Polar bear	Arctic	8000
Blue whale	Oceans	6000
Orangutan	Borneo	5000
Black rhino	Africa	3000
Asiatic buffalo	India	2000
Indian rhino	India	600
Monkey-eating eagle	Philippines	300
Florida panther	Florida, USA	300
Lion tamarin	South America	300
Siberian tiger	Russia, China	200
Giant panda	China	200
Imperial eagle	Spain	100
Java rhino	Indonesia	50
Andean condor	South America	26

PLANTS

PLANTS

● Plants make their own food from the simplest ingredients. Green plants take in carbon dioxide from the air, and water and minerals from the ground. With the energy from sunlight they make them into very complex substances such as proteins and sugars. Plants also release oxygen into the air. There are almost 300,000 species of flowering plants and about 150,000 of all other plant species.

● Fungi do not absorb sunlight but use animals and plants, both dead and living, as their source of food. They are usually classified separately from plants because of this.

MEAT EATERS

DID YOU KNOW?

About 500 different types of plant eat animals.

● The color and scent of the **Venus fly trap** attract insects. As soon as an insect touches hairs on the leaves the two halves snap shut, trapping the

A young frog trapped by a Venus fly trap.

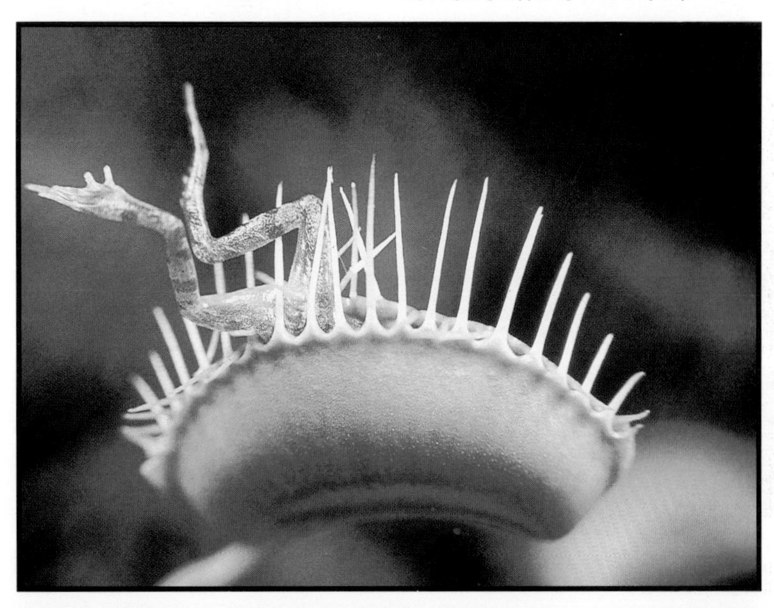

insect. The plant then produces a liquid which dissolves the insect. It takes about 10 days to completely digest a fly, and then the leaves open again ready for their next victim.

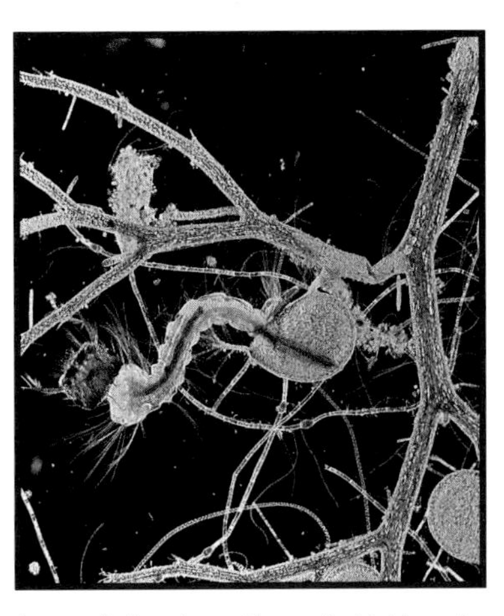

A mosquito larva trapped by greater bladderwort.

● **Pitcher plants** can grow large enough to trap a mouse, lizard or frog. These are attracted to the plant to drink, as the pitcher looks like it is full of water. But the pitcher plants main source of food is insects, which are attracted by the nectar around the rim of the pitcher. Once the animal moves over the rim it slides down to the bottom of the pitcher and into a pool of digestive juices, produced by the plant. It is unable to crawl out and is slowly digested. The hood on the pitcher plant acts as an umbrella to stop the plant filling with rain.

Tropical pitcher plant. Trapped insects provide the plant with nitrogen so that it can survive in nitrogen poor soil.

● **Bladderworts** trap tiny creatures in small sacs or bladders, that they grow underwater. The larger bladderworts can even trap small fish. When a creature touches hairs on the bladder entrance, an inward opening door flies open sucking in water and the helpless creature. The bladder has a special system for removing water, and the door is then able to close again leaving the victim trapped in the bladder and awaiting digestion.

TREES

● The largest living thing on Earth is a **giant sequoia**, a coast redwood, growing in California and given the name 'General Sherman'. It is 84 m (275 ft) tall and about 31 m (103 ft) round its trunk. It weighs over 2000 tons and that is equal to the weight of almost 700 fully grown African elephants. This tree is at least 2,200 years old. The tallest tree is a Californian **redwood** which measures about 110 m (360 ft) tall. This is about the same height as some of the largest pyramids in Egypt.
The oldest living tree in the world is a **bristlecone pine** growing in California. It has been given the name 'Methuselah' and is about 4,600 years old.

● The fastest growing tree is the Malaysian *Albizzia falcata*. It can grow as much as 10 m (33 ft) in one year. Some types of **eucalyptus** can also grow at about this rate.

● One of the slowest growing trees is the **sitka spruce** found in Arctic regions. It only grows about 28 cm (11 in) in a hundred years.

The largest living thing: a giant sequoia named General Sherman. Sequoias are conifers which grow in the mountains of California.

● The largest leaves in the world grow on the ***Raphia ruffia palm***. They can measure up to 20 m (66 ft).

● The **banyan** tree sends down roots from its branches. In time these roots become thicker and support the tree. They also produce their own branches which send down more roots. In India a single banyan tree has produced a jungle of roots and branches over 600 m (2000 ft) in circumference. It would take about 10 minutes to walk round it; it takes about 10 seconds to walk round a big tree in your local park.

● An enormous amount of water is needed by a tree every day. It passes through the roots, up the tree and evaporates from the leaves. As much as 900 litres (200 gallons) pass up and out of an average-sized **oak** tree every day. If you turned a cold water tap full on in the sink it would have to run for about an hour to provide this much water.

DID YOU KNOW ?

The coco-de-mer palm tree which grows in the Seychelles, produces the largest seeds of any plant. They can weigh up to 24 kg (50 lb).

A huge banyan tree at Ranthambhor National Park, Rajasthan, India. The branches have roots that grow down, take root, and begin new trunks. One banyan tree may have hundreds of trunks.

A giraffe browsing on an acacia tree. The very thick saliva of the giraffe provides some protection against the acacia's thorns.

● The **bull's horn acacia** tree not only has spines to protect it from grazing animals, but it is also swarming with ants. The ants live in the spines and feed on sugar produced by the tree. Any animal which disturbs the ants is vigorously attacked and cannot feed on the tree.

MORE AMAZING PLANTS

● **Lianas** are climbing plants that grow up through the trees of tropical rainforests to reach sunlight. Their long woody stems hang down like ropes and can reach 300 m (1000 ft) in length.

● ***Rafflesia*** has the largest and probably smelliest flower in the world. It grows in the forests of south east Asia and has a huge flower up to 90 cm (3 ft) across, which lasts about a week. A single bloom can weigh 7 kg (15 lb). It is an ugly flower covered in warts and smelling like rotting flesh. Flies attracted by the smell, swarm over the flower and pollinate it. *Rafflesia* is a parasite. It has no leaves and cannot make its own food. It grows on the roots of lianas and takes all its food from them.

● Many tropical **orchids** grow on the branches of trees and have aerial roots which absorb water from the moist air

Forest with lianas, Pantanal, Brazil. Lianas are vines found chiefly in tropical rainforests.

Rafflesia is a parasitic flower that flowers once every ten years. It is the world's largest flower.

around them. One unusual orchid, ***Vriesa***, has tiny scales on its leaves, in the shape of little cups, which trap and absorb water.

● The most seeds in a single capsule are found in orchids. Some species can have as many as 20,000. These tiny seeds are so light that 3 million of them only weigh one gram (0·03 oz).

● One of the smallest flowering plants is ***Wolffia arrhiza***. It is about 5 mm (0·02 in) across and is found floating in ponds.

● Leaves of the **giant water lily** can grow up to 2.5 m (8 ft) across.

Water lilies grow in shallow water on stalks rooted in the mud beneath.

Welwitschia mirabilis in the Namib Desert, south west Africa

● Despite its appearance of several ragged leaves the ***Welwitschia mirabilis*** only produces two leaves throughout its life, although it may live for more than 100 years. Each leaf grows about 5 cm (2 in) a year. The largest leaves on record measured about 8 m (26 ft) long.

● Insects attracted to the **cuckoo pint's** foul smell slip down and are trapped inside the flower. They can only escape when the flower begins to die, and they leave covered in pollen to pass on to other flowers.

● The flowers of **figs** are in the center of the fig fruit. Special fig wasps carry pollen from the male to female flower.

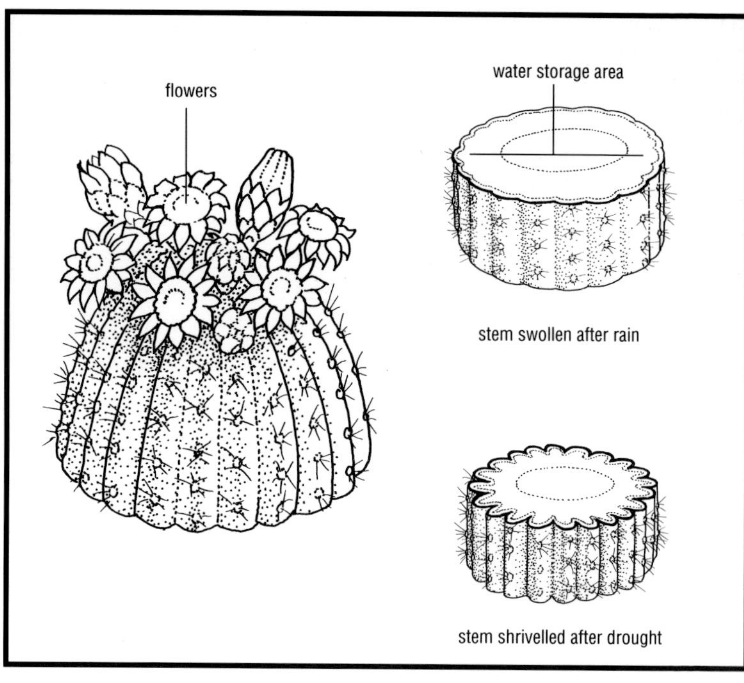

During the drought the stem of a giant saguaro cactus shrinks and develops deep pleats. But after heavy rain it swells, the stem expands and the pleats almost disappear.

● Scientists have found that a **rye** plant can produce 620 km (385 miles) of roots in 4 months; that is about 5 km (3 miles) a day!

● The giant **saguaro cactus**, found in the Arizona desert, grows to over 15 m (50 ft) high and may live for over 200 years. Its roots extend over an area as much as 100 m (330 ft) from the stem.

● Cacti have many tricks to help them survive in harsh desert climates. The surface of a cactus is coated with a waxy layer that prevents water evaporating from the plant. The small pores through which the cactus breathes are kept tightly closed during the heat of the day, and only open at night to take in carbon dioxide and release oxygen. It is able to store large quantities of water, and its prickles protect it from animals that might want to feed on its juicy flesh. Some cacti have vicious spines more than 15 cm (6 in) long. The spines also help in providing the cactus with water. In morning mists tiny drops of water form on the spines, in the way dew forms on grass. The drops then drip down to the ground for the roots to absorb.

- In the swirling mists on the mountains of Kenya grow monstrous plants. Normally plants such as **lobelia** and **groundsels** are under 3 feet high but in this region they grow several yards high. The reason for this spectacular growth is unknown.

DID YOU KNOW?

The tallest grasses in the world are the thorny bamboos of India, which can grow as high as 37 m (120 ft).

- Some species of **bamboo** only flower every 60 years. Others flower after 20 years and then die.

- You can almost watch a bamboo grow. Some grow up to 3 ft a day. That is on average 1½ inches every hour.

- The longest seaweed is the Pacific **giant kelp** which can grow to over 60 m (200 ft) long.

A forest of kelp beneath the sea.

FUNGI

● Fungi do not have green leaves so they cannot photosynthesize and use the energy of sunlight to make the chemicals they need to survive. Instead they directly absorb these substances from the plants or animals they live on. The fungus grows a spreading network of fine threads, called mycelium, through its food source, and this extracts all that the fungus needs.

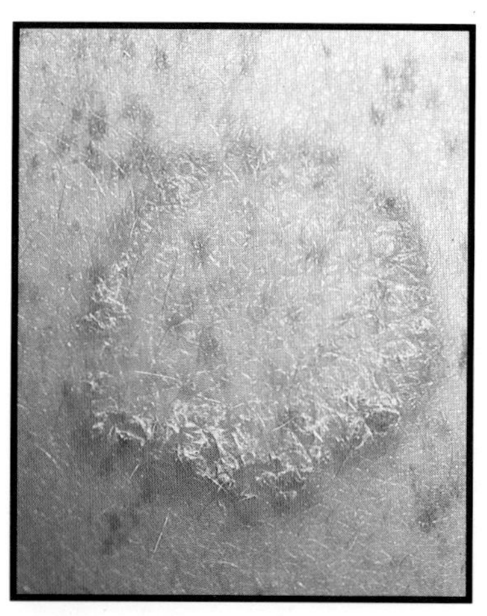

Ringworm on a human arm. Ringworm is a fungal infection of the skin. It is highly infectious.

● Some fungi feed on living creatures such as caterpillars and beetles. Eventually the fungus spreads over the animal and kills it.

● A fungus which feeds on eelworms, which are found in the earth, captures them alive in sticky nets which tighten and strangle the eelworm as it struggles to escape.

● Some fungi feed on you! **Athlete's foot**, **thrush** and **ringworm** are caused by fungi which feed on our skin.

● The **giant puffball** fungus can grow to about 30 cm (12 in) across and produces seven trillion (7,000,000,000,000) spores.

DID YOU KNOW ?

The kerosene fungus (*Amorphotheca resinae*) can live in jet fuel tanks. If there is a tiny amount of water in the tank the fungus can use the fuel as food.

● The **stinkhorn** fungus is very fast growing. It can grow at the rate of 5 mm (0·2 in) a minute, and can reach its full size in 20 minutes. Its smell of rotting flesh attracts flies, which crawl all over it and collect the spores, scattering them on their travels.

The stinkhorn fungus attracts flies because it smells of rotting flesh.

● The Japanese fungus *Mycaena lux-coeli* glows in the dark and can be seen as far away as 15 m (50 ft). It is not known why this and, some other fungi, produce light.

A boy holding a giant puffball found in Lincolnshire, England. **Warning**: most fungi are poisonous, so do not touch.

LICHENS

● Lichens look like a single plant, but are in fact an alga and a fungus living together. The cells of the algae are held together by strands of fungus. The algae are green and produce sugars through photosynthesis. The fungus feeds on these sugars and provides the algae with the water it needs. Lichens thrive in places no other plants can possibly live. They are found on barren rocks in deserts and even in the Arctic and Antarctic.

● Some lichens feed on rocks. They dissolve the rock with acid and absorb the minerals through the roots.

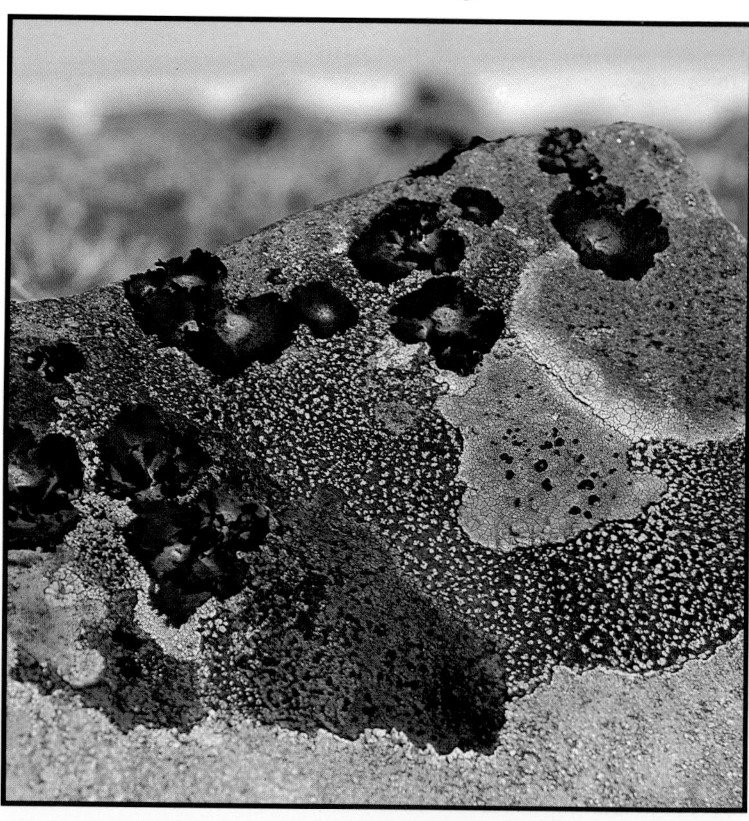

Four different lichen varieties growing on rock. Lichen is a good indicator of the level of pollution as some lichens are very sensitive, and cannot grow where air pollution is above a certain level.

Photo credits

Heather Angel/Biofotos pp 19, 65;

Ardea pp 63 (bottom), 135/A Warren p 105;

Biophoto Associates p 6;

Bubbles/F Rombout p 11;

Bruce Coleman/J Van de Kam p 58 /J Van Wormer p 63 (top) /J Burton pp 64 (top), 91, 113 /L Lyon p 68 /N Myers p 71 /F Futil p 73 (bottom) /K Wother p 77 (bottom) /L Marigo p 78 /H Albrecht p 81 /J Markham p 82 /J Cancalosi p 86 /A Power p 87 /M Fogden p 90 /pp 99, 100 (left) /G Cubitt pp 102, 132 /p 121 /D Austen p 130 /F Sauer p 134 /pp 143 (top), 145, 147 (all), 148, 153 (top), 154;

Frank Lane Picture Agency pp 64 (bottom) 101;

Oxford Scientific Films/A MacEwen pp 53, 54 (left) /J Cooke pp 54 (right) 133 /Animals Animals p 57 /Z Leszczynski p 59 /D Guravich p 72 /S Turner p 73 (top) /K Westerskov p 74 /T McCann p 77 (top) /R Davies p 80 /Partridge Films p 84 /P Ryley p 89 /p 92 /M Linley p 94 /C Houghton p 99 (bottom) / M Birkhead p 100 (right) /M Fogden pp 104, 137 /K Atkinson pp 110, 118 /F Bavendam pp 111, 119, 123, 124 /pp 107, 116, 120, 122 /G Bernard p 125 /E Robinson p 126 /D Clyne p 129 /S Camazine p 131 /S Morris p 136 /pp 135, 139, 141, 142, 143 (bottom), 144, 146, 149, 151, 153 (bottom);

Planet Earth pp 103, 112;

Science Photo Library pp 23, 152.

Artwork credits

Graham Allen (Linden Artists): 55, 61, 67, 70, 75, 79
Norman Arlott: 96, 97
Brian Beckett: 30, 50, 95, 127, 138
Peter Joyce: 47
Frank Kennard: 19, 21, 25, 28, 29, 33, 34, 37, 40, 42, 44
Mick Loates (Linden Artists): 108-109, 114
Annabel Milne: 5, 10, 14
Oxford Illustrators: 7
Jim Robbins: 35
Mike Saunders: 8, 12, 24, 32, 39, 46, 48
Michael Woods: 12, 85, 88, 150
W. D. Phillips & Nick Hawkin Associates: 31